T0192474

A Practical Guide to Effective Workplace Accident Investigation

A Practical Guide to Effective Workplace Accident Investigation

Ron C. McKinnon

CRC Press
Taylor & Francis Group
Boca Raton London New York

CRC Press is an imprint of the
Taylor & Francis Group, an **informa** business

First edition published 2022
by CRC Press
6000 Broken Sound Parkway NW, Suite 300, Boca Raton, FL 33487-2742

and by CRC Press
4 Park Square, Milton Park, Abingdon, Oxon, OX14 4RN

CRC Press is an imprint of Taylor & Francis Group, LLC

Reasonable efforts have been made to publish reliable data and information, but the author and publisher cannot assume responsibility for the validity of all materials or the consequences of their use. The authors and publishers have attempted to trace the copyright holders of all material reproduced in this publication and apologize to copyright holders if permission to publish in this form has not been obtained. If any copyright material has not been acknowledged, please write and let us know so we may rectify in any future reprint.

Library of Congress Cataloging-in-Publication Data
A catalog record has been requested for this book

ISBN: 978-1-032-05444-5 (hbk)
ISBN: 978-1-032-11488-0 (pbk)
ISBN: 978-1-003-22009-1 (ebk)

DOI: 10.1201/9781003220091

Typeset in Times New Roman
by codeMantra

Contents

SECTION II   Accident Investigation Methodology

SECTION III   Investigating and Analyzing the Event

SECTION IV   Remedial Measures to Prevent a Recurrence

SECTION V   *Evaluating the Quality of Accident Investigation Reports*

SECTION VI   Accident Scenarios

SECTION VII   Conclusion

About the Author

Ron C. McKinnon, CSP (1999–2016), is an internationally experienced and acknowledged safety professional, author, motivator, and presenter. He has been extensively involved in safety research concerning the cause, effect, and control of accidental loss, near-miss incident reporting, accident investigation, safety promotion, and the implementation of health and safety management systems for the last 46 years.

The author received a National Diploma in Technical Teaching from the Pretoria College for Advanced Technical Education, a Diploma in Safety Management from the Technikon SA, South Africa, and a Management Development Diploma (MDP) from the University of South Africa, in Pretoria. He received a Master's Degree in Health and Safety Engineering from the Columbia Southern University.

From 1973 to 1994, Ron C. McKinnon worked at the National Occupational Safety Association of South Africa (NOSA), in various capacities, including General Manager of Operations and then General Manager Marketing. He is experienced in the implementation of health and safety management systems (SMS), auditing, near-miss incident and accident investigation, and safety culture change interventions.

From 1995 to 1999, Ron C. McKinnon was a safety consultant and safety advisor to Magma Copper and BHP Copper North America, respectively. In 2001, Ron spent two years in Zambia introducing the world's best safety practices to the copper mining industry. After leaving Zambia, he was recruited to assist in the implementation of the world's best class safety management system at ALBA in the Kingdom of Bahrain.

After spending two years in Hawaii at the Gemini Observatory, he returned to South Africa. Thereafter, he contracted as the Principal Health and Safety Consultant to Saudi Electricity Company (SEC), Riyadh, Saudi Arabia, to implement a world's best practice safety management system, throughout its operations across the Kingdom involving 33,000 employees, 27,000 contractors, 9 consultants, and 70 Safety Engineers.

Preface

Accident investigation can only be effective if it examines all the facts of the event in an unbiased manner and proposes remedial measures that can be implemented, which in turn can prevent the recurrence of similar events. This book provides a practical guide to the effective investigation of high-potential near-miss incidents, injury, and property-causing accidents.

Any accident investigation process that does not fix the problem is a waste of time. Many investigations turn into witch-hunts and want to find and punish a guilty party or parties. This will not fix the problem. The book focuses on identifying and fixing the problem and is centered on the safety management principles of the *critical few*, the principle of *definition*, and the principle of *multiple causes*.

The principle of the *critical few* states that "*a small number of basic causes could give rise to the majority of safety problems.*" This means a few critical jobs could be responsible for the majority of accidents occurring, and these few critical jobs should receive maximum safety control to minimize their potential for causing (the majority) of accidents.

This principle of *safety definition* states that "*decisions concerning the remedial measures taken after an accident can only be made if the basic (root) causes of loss producing events are clearly identified.*" This means that a logical and proper decision can be made only when the root causes or real problems are first defined. Prescription without diagnosis is malpractice.

The principle of *multiple causes* states that "*accidents, near-miss incidents, and other problems are seldom, if ever, the result of a single cause.*" If the accident investigation system is not structured and does not follow the loss causation sequence, and determine both the immediate and root causes of the event, the system is basically worthless.

Many accident investigators fear delving deeper into the causes of the event as this may open a can of worms. An effective accident investigation may well open a can of worms. This book emphasizes that the investigation must be a fact-finding mission and not a fault-finding exercise.

To be effective and contribute toward the prevention of future accidents, the organization should declare amnesty and create an environment within which employees can report near-miss incidents and accidents confidently, and where investigations can delve into all the causes, irrespective of what will be uncovered.

Acknowledgments

This publication is based on many years of being involved in accident and near-miss investigations and of teaching others to be effective accident investigators.

Much of the knowledge and information in this book was gained from the people and organizations that I have associated with and worked with during my safety career. I thank them and my mentors for sharing their knowledge and experience with me. It was a pleasure knowing and working with them. I also pay tribute to the safety professionals who I have quoted in this book. They need to be thanked for their diligent research into safety management, and especially accident investigation.

For making this publication possible, I thank my wife, Maureen McKinnon, who spent numerous weeks editing this manuscript. This support warrants my deep gratitude.

NOTE: Every effort has been made to trace rights holders of quoted passages and researched material, but if any have been inadvertently overlooked, the publishers would be pleased to make the necessary arrangements at the first opportunity.

Section I

Introduction to Accident Investigation

1 Introduction

OBJECTIVE

The objective of this book is to provide the reader with the knowledge and understanding of a practical and efficient method with which to investigate workplace accidents and high-potential near-miss incidents. It is also intended to enable the reader to identify all the contributing causes of the event, so that positive remedial measures can be taken, to prevent a recurrence of a similar event in the future.

WORKPLACE INJURY AND DEATH STATISTICS

According to the International Labor Organization (ILO) (2021):

> The ILO estimates that some 2.3 million women and men around the world succumb to work-related accidents or diseases every year; this corresponds to over 6,000 deaths every single day. Worldwide, there are around 340 million occupational accidents and 160 million victims of work-related illnesses annually. The ILO updates these estimates at intervals, and the updates indicate an increase of accidents and ill health.

> *(ILO website, Copyright © International Labor Organization 2021)*

The National Safety Council (US) *Injury Facts 2019*, (2019a), reports that the US experienced 4,572 work-related fatalities and 4.64 million medically consulted workplace injuries (NSC Injury Facts 2019).

According to the Health and Safety Executive (HSE) UK (2019), 1.6 million working people suffering from a work-related illness, 111 workers killed at work, and 693,000 working people sustain an injury at work according to the Labor Force Survey. This led to 38.8 million working days lost due to work-related illness and workplace injury (p.8).

On reviewing these injury numbers from around the world, it is clear that millions of accident investigations are carried out each year in an effort to determine the causes and put preventative measures in place. If accidents are investigated thoroughly and remedial control measures are put in place to prevent recurrences, the injury and fatality tolls should be reduced.

CONCEPTS EXPLAINED

To ensure the reader is familiar with occupational health, safety, and accident investigation terminology which will be used throughout this book, some of the main concepts are defined and explained in this chapter.

DOI: 10.1201/9781003220091-2

SAFETY

Safety is the control of all forms of accidental loss that could occur in a workplace, normally as a result of accidental, undesired, unplanned, or unexpected downgrading events occurring. These losses could include injury to employees, occupational diseases, damage to equipment and property, or business interruption. These are all undesired events.

Safety has also been defined as being safe from harm, either acute or chronic. Workplace safety concerns the health, safety, and well-being of employees in the environment in which they work. Safety also encompasses the prevention of damage to equipment and products.

Other areas of health and safety are health and safety in the home; recreational safety; sports safety and road safety, to mention a few. This book focuses on investigating accidents and near-miss incidents that occur at workplaces, or during work-related activities outside the workplace.

PROACTIVE SAFETY VERSUS REACTIVE SAFETY

Most of the controls within the safety management system (SMS) are proactive processes, but some necessary reactive control activities are also included. An accident investigation is a reactive activity, but does impact on the proactive side as it recommends upstream activities to prevent the recurrence of similar events downstream.

OCCUPATIONAL HYGIENE

Occupational hygiene is the science devoted to the anticipation, recognition, identification, evaluation, and control of environmental stresses arising within a workplace, which may cause illness, impaired well-being, discomfort, and inefficiency of employees or others.

Occupational hygiene is also described as the process of dealing with the influence of the work environment on the health of employees. The objective of occupational hygiene is to recognize occupational health hazards, evaluate the severity of these hazards, and eliminate them by instituting effective control measures. Accidental exposures should be investigated with the same rigor as acute injuries.

ACCIDENT

An accident is an undesired event that results in a loss. The loss may be

- Injury
- Fatality
- Multiple fatalities
- Work-related disease
- Damage
- Process/product loss

- Fire loss
- Explosion
- Environmental damage or pollution

In many cases, the accidental event leads to a multitude of losses including some listed above.

INJURY

An injury is the bodily harm to a person sustained as a result of an accidental contact with a source of energy, which is greater than the body's threshold limit. This includes any illness or disease arising out of, and during the course of normal employment.

ACUTE VERSUS CHRONIC

The item that inflicts the injury or illness is the agency that could be an occupational hygiene agency or a general agency. Injuries caused by accidents are normally immediate (*acute*). Industrial diseases are mostly long term (*chronic*) as they manifest over a period of time. The exchange of energy in diseases is normally referred to as an exposure. There is an exchange of energy the same as in an injury-related accident, except it is phased over a longer time.

INJURY VERSUS ACCIDENT

Unfortunately, the term *injury* has over the years become synonymous with the term *accident*. Many get the two concepts confused and often an injury is called an accident. An injury is physical harm to the human body and may be acute or chronic. An accident is the sequence of events that culminate in unintentional injury or illness.

INJURY SEVERITY LUCK FACTOR

The severity of an injury or disease is a measure of how serious the injury or disease is. It is sometimes measured by the workdays lost as a result of the injury or disease. The *Luck Factor* theory proposed by McKinnon (2000) suggests that fortuity often determines the severity of an injury in the accident sequence.

UNINTENDED LOSS

Any accident, or undesired event, leads to a loss of some form or another. These losses are not budgeted for and create additional burdens to the organization, which are straddled with injured employees, damage to equipment or property, and possible production delays caused by events that seemingly "just happen." These undesired and unplanned losses are *caused* and do not *just happen*. They are caused by undesired events called accidents and can be prevented.

BUSINESS INTERRUPTION

Accidental events could also cause interruptions to the business proceedings. An injury-causing accident could slow the business at hand down, while the injured person receives attention and first aid treatment. Stopping of processes or machinery and cleaning up after an accident are all factors that lead to disrupting the day-to-day business. Damage to equipment or products caused by accidents also results in disruptions. Often, if there is no injury as a result of an accident many do not rate it as important, and the interruption and cost of damage are overlooked and not viewed as accidental events. These types of accidents should be investigated.

LOSS OF PRODUCT AND PROCESS

Accidental spills or delays caused by accidents can impact the production and create a loss. After an injury in the workplace, there is a disruption to the workflow and process. Employees are taken away from their workstations to attend to the injured person. The site may even be shut down for the investigation, and in some instances, machines and production lines may be stopped.

ENVIRONMENTAL HARM

Accidental discharge or spill of toxic and other polluting materials and substances may have a detrimental effect on the environment. Although there may be no injury to employees, there may be harm to the environment. This harm could be in the form of water, air, soil pollution, or poisoning.

ACCIDENT VERSUS INCIDENT

Many confuse the terms *accident* and *incident*. To simplify, an *accident* is an event that causes a loss, and a *near-miss incident* is an event that *could* have caused a loss, under different circumstances. The term *incident* is used to describe all downgrading events and refers to events that happened or could have happened. The terms that will be used in this book are *accident* (an undesired event that results in a loss), and a *near-miss incident* (an undesired event that does not result in loss, but which had the potential to cause loss).

NEAR-MISS INCIDENTS

Near-miss incidents are also known as near-misses, near hits, incidents, close shaves, warnings, or near hits. Other familiar terms for these events are close calls, or in the case of moving objects, near collisions.

Near-miss incidents are near-miss events that come close to causing some form of loss. There was an actual flow of energy, but no contact occurred. In some instances, the flow of energy may have dissipated without making any contact, thus causing no loss. In most cases, the energy does not make contact with anything, or anybody, thus causing no harm. In some cases, the exchange of energy during the contact was insufficient to cause loss or injury. The fact that there was an undesired exchange of energy is reason enough to heed the warning. Near-miss incidents are accidents waiting in the shadows.

Precursors to Accidents

In the workplace, this means that a serious loss-causing accident almost occurred. Near-miss incidents are the precursors to accidents (which lead to an exchange of energy and subsequent loss). They are warnings. Literally, fractions of an inch, or a split second, maybe the difference between a serious loss-producing accident and a near-miss incident.

Potential for Loss

Not all near-miss incidents should be subjected to a full accident investigation process. All should be reported and risk assessed. Those with high potential for loss should be treated as accidents and be investigated in the same way as accidents are. Effective high-potential near-miss investigations are a proactive activity that helps find and eliminate the source of the event before a loss-producing accident occurs.

Near-Miss Incident Luck Factor

McKinnon (2000) further proposes that the difference between a high-potential near-miss incident and a loss-producing accident is often determined by luck (p.124).

The HSE UK (2004a) supports this theory:

> It is often pure luck that determines whether an undesired circumstance translates into a near miss or accident. The value of investigating each adverse event is the same.
>
> Adverse events have many causes. What may appear to be bad luck (being in the wrong place at the wrong time) can, on analysis, be seen as a chain of failures and errors that lead almost inevitably to the adverse event. This is often known as the Domino effect. (p.6)

Risk Matrix

A *risk* is defined as, *any probability, likelihood, or chance of an accidental loss*. A simple way of assessing the risk of a near-miss incident is to rank the happening on a risk matrix, which will give an indication of what *could* have happened under different circumstances (Figure 1.1).

HAZARD IDENTIFICATION AND RISK ASSESSMENT (HIRA)

The purpose of HIRA is to identify hazards and evaluate the risk of injury or illness arising from exposure to a hazard, with the goal of eliminating the risk, or using control measures to reduce the risk at the workplace.

The organization should perform ongoing documented HIRAs for the work at hand. The assessments should identify competencies required of the employees, as well as controls and barriers required to guard against identified hazards. The assessments should include identifying which company policies, standards, procedures, and processes apply to the workplace situations. The organization's health and safety department may be consulted to establish a risk ranking of the various work processes.

A HIRA process includes

- Identifying tasks done, before, during, and after the work,
- Identifying potential hazards for each task,
- Determining the level of risk, using probability and consequence on the risk matrix,
- Identifying controls to be put in place, with consideration given to how effective and adequate the proposed controls would be.

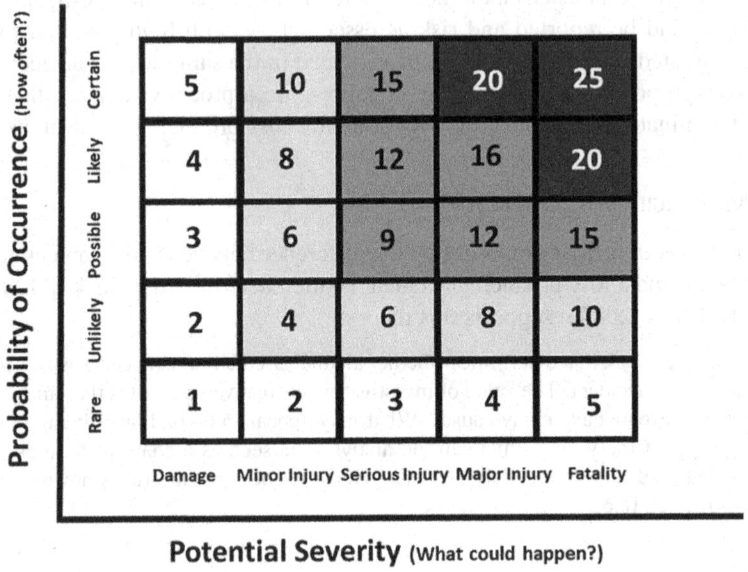

FIGURE 1.1   A simple risk matrix.

HIERARCHY OF CONTROL

There should be multiple layers of controls protecting employees from hazards and associated risks. The types of controls could include

- Elimination – which would mean changing the way the work is to be done.
- Engineering controls – such as the re-design of a worksite, equipment modification, and tool re-design.
- Administrative controls – such as a change of work methods and re-scheduling of work.
- Health and safety system controls – such as health and safety policies and standards, work procedures, training, and personal protective equipment (SMS elements).

An effective accident investigation will recommend one or more or a combination of these remedies to prevent a similar accident occurring in the future.

SMS CONTROLS

The organization's SMS would consist of a number of control mechanisms, which are aimed at identifying hazards and reducing their risk as far as is reasonably practicable. Work standards, procedures, and policies within the SMS require an ongoing and proactive application to reduce the probability of an accident occurring. One of the processes within the SMS would be near-miss incident and accident investigation requirements and protocols.

ACCIDENT CAUSATION

The HSE UK (2004b) proposes three layers of accident causes. These causes are classified as:

- Immediate causes – the agent of injury or ill health (the blade, the substance, the dust, etc.)
- Underlying causes – unsafe acts and unsafe conditions (the guard removed, the ventilation switched off, etc.)
- Root causes – the failure from which all other failings grow, often remote in time and space from the adverse event (failure to identify training needs and assess competence, low priority given to risk assessment, etc.) (p.6).

For purposes of simplicity, this publication will use *immediate* and *root* causes for the purpose of an accident investigation, as the above-mentioned *immediate causes* commonly refer to the agency or agency part which will be discussed in the later chapters.

ACCIDENT IMMEDIATE CAUSES

The immediate causes of accidents, sometimes referred to as *proximate* causes, are the high-risk behaviors and high-risk workplace conditions, which lead directly to the exchange of energy and the loss.

HIGH-RISK CONDITION (UNSAFE CONDITION)

A high-risk workplace condition is any physical or environmental condition that constitutes a hazard and may lead to an accident if not rectified.

HIGH-RISK BEHAVIOR (UNSAFE ACT)

High-risk behavior is a departure from a normal accepted or correct work procedure, which reduces the degree of safety of that procedure.

ACCIDENT ROOT CAUSES

Root causes, sometimes called *underlying* causes, are the most fundamental causes that can be reasonably corrected to prevent recurrence of the error. The root causes are never the same as the immediate causes, and the root cause is the management system problem that allowed the high-risk behavior and conditions to happen or to exist. They are the underlying or systemic, rather than the generalized or immediate causes.

The two categories of accident root causes that will be used in this book are the Personal (human) factors and the Job (workplace and environment) factors.

ACTS OF NATURE

Another category of accident root causes are acts of nature such as flash floods, volcanic eruptions, and earthquakes, which are beyond our control.

The HSE UK (2004c) define the root cause as

> An initiating event or failing, from which all other causes or failings, spring. Root causes are generally management, planning or organizational failings.

> *(p.5)*

PERSONAL (HUMAN) FACTORS

The personal factors could be factors such as stress, lack of skill, lack of knowledge, inadequate motivation, physical or mental shortcomings, or a poor attitude toward health and safety that culminate in high-risk behavior.

JOB (WORKPLACE AND ENVIRONMENT) FACTORS

Job factors could include items such as excessive wear and tear, inadequate health and safety standards, tools and equipment that are insufficient, poor or not maintained, no standards for purchasing, lack of maintenance, or poor supervision.

ACCIDENT AND NEAR-MISS INCIDENT INVESTIGATION

The purpose of investigating accidents is to determine what happened, why it happened, what the consequences were, what the immediate and root causes were, and what action needs to be taken to prevent a recurrence of a similar event in the future. In most instances, investigating accidents at the workplace is a legal requirement as well.

According to Safeopedia,

> Accident investigation is a process of systematic collection and analysis of information relating to an accident that led to the loss of property, time or health of individuals, or even the loss of lives.
>
> Accident investigation is the scientific and academic analysis of the facts that occurred during an accident. An investigation is conducted to identify the root cause of an accident in an effort to make recommendations or take corrective actions to prevent the future occurrence of the same or a similar event (Website) (2020).

Effective Accident Investigation

An accident investigation is only effective when it is done thoroughly, and the root causes of the event are identified and systems put in place to prevent a recurrence. Merely going through the motions of the investigation, or finding a guilty party, does not solve the problem and is therefore ineffective. The management principle of definition states that a problem, such as an accident, can only be solved once the real causes have been identified.

Near-Miss Incident Investigation

High-potential near-miss incidents should be regarded as serious as loss-producing accidents and should be investigated using the same methodology and diligence as for accidents. The term *accident investigation* in this publication, therefore, refers to the investigation of accidental events that cause injury, ill health or occupational disease, property damage and business interruption, as well as high-potential near-miss incidents.

CONCLUSION

Workplace hazards should be identified by inspection and risk assessment techniques, and suitable controls should be put in place to protect workers from these hazards. Should hazard control methods fail, a thorough investigation should be launched to ascertain the immediate and root causes so that preventative measures can be put in place.

All undesired and unplanned downgrading events, such as accidents, property damage, and high-potential near-miss incidents, are caused. They simply do not just happen. Since they are caused, they can be prevented. They should be investigated thoroughly and the causes identified so that action can be taken to prevent a recurrence of the same or a similar event in the future.

2 The Philosophy of Safety

WORKPLACE HEALTH AND SAFETY

Health and safety in the workplace is primarily about keeping employees free from any form of harm arising from the workplace. The employer's obligation is to provide a clean, neat, safe, and healthy work environment for employees. This is so that they can return home after work in the same condition as they were when they came to work.

The safety of employees was not always a priority until laws and regulations were promulgated which called for certain actions to protect employees at work. Devastating workplace accidents that killed hundreds of workers helped bring about more stringent regulations to protect workers. Early safety legislation was drafted mainly to protect orphaned children who were being exploited by being forced to work in cotton mills and other industries.

Today, the health and safety of employees has become an international concern and more and more organizations are embracing it as part of their business culture.

MANAGEMENT FUNCTION

Although it is generally accepted that safety is a function of management, this is rarely practiced, especially after an accident, where traditionally the blame is put on the shoulders of the employee. One of the safety's biggest paradigms is that employees are to blame for accidents. This paradigm has to change before an accident investigation can be effective. If not, the investigation becomes a blame game, which is ineffective and does not solve the problem.

SAFETY AUTHORITY, RESPONSIBILITY, AND ACCOUNTABILITY

Occupational health and safety starts at the top, is supported by the top, and is led by top management. The success or failure of the organization's safety efforts falls directly on the shoulders of the management team. Each manager's level of safety authority, safety responsibility, and safety accountability should be clearly defined and understood.

SAFETY AUTHORITY

Safety authority is a formal, specified authority, which gives a manager the authority to act on behalf of the organization. These powers of authority are normally spelt out in the manager's job description. All safety management system (SMS) standards should include responsibility and accountability clauses.

DOI: 10.1201/9781003220091-3

SAFETY RESPONSIBILITY

Safety responsibility is the obligation entrusted to the possessor, for the proper custody, care, and safekeeping of property or supervision of individuals. With responsibility goes authority to direct and take the necessary action to ensure success.

SAFETY ACCOUNTABILITY

Safety accountability is when a manager is under obligation to ensure that safety responsibility and authority are used to achieve health and safety standards. It means being liable to be called on to render an account, or be answerable for decisions made, actions taken, outcomes, and conduct.

ACCIDENT INVESTIGATION RESPONSIBILITY

Since managers have the authority and responsibility for health and safety, they also have the obligation to lead accident investigations and be active participants in the process. Irrespective of the accident causes, they are accountable to ensure the investigation is done correctly and that recommended controls are implemented. Delegating this down to the health and safety department is not acceptable. Only the management has the authority and ability to create a safe and healthy workplace.

MANAGEMENT LEADERSHIP

According to the Occupational Safety and Health Administration (OSHA),

> Management provides the leadership, vision, and resources needed to implement an effective safety and health program. Management leadership means that business owners, managers, and supervisors:
>
> - Make worker safety and health a core organizational value.
> - Are fully committed to eliminating hazards, protecting workers, and continuously improving workplace safety and health.
> - Provide sufficient resources to implement and maintain the safety and health program.
> - Visibly demonstrate and communicate their safety and health commitment to workers and others.
> - Set an example through their own actions (OSHA 2021).

UPSTREAM HEALTH AND SAFETY MANAGEMENT SYSTEMS

Upstream systems and controls are the processes and procedures that an organization has in place to prevent accidents occurring. These are incorporated in the health and safety management system. They occur on an ongoing basis and are part of the day-to-day management of the organization. Many organizations react after the event, which is ineffective as the loss has already occurred.

PRE-CONTACT

This is the phase before the accident. It involves the systems, standards, and actions that are in place before any accidental contact with an energy source takes place. Examples are as follows:

- Providing guards and barriers
- Good stacking and storage practices
- Safety awareness training
- Safe work procedures
- Other controls within the SMS.

CONTACT

This is the phase in the accident sequence where the injury, damage, or other loss is created. It is where the accidental contact with an energy source takes place. Previously, this was incorrectly termed the *accident*, but in fact it is the segment of the accident where the loss occurs. It is where the blade cuts into a finger, where the acid enters an eye, where a falling body strikes the ground, or where harmful fumes are inhaled into the lungs.

CONTACT CONTROL

There is little opportunity for safety control at this stage in the accident sequence. The only control that is effective at the contact phase is the wearing of personal protective equipment (PPE). The PPE does not stop the flow of energy, but reduces the consequence of the energy exchange. This contact control is deemed as the last resort and is only reverted to when all pre-contact efforts have been exhausted.

POST CONTACT

An accident investigation is a post-contact activity. Near-miss incident reporting and investigation is a pre-contact activity. This is why it is so important to investigate high-potential near-miss incidents. Many organizations make the mistake of trying to investigate all near-miss incidents. This is a waste of effort. Reported near-miss incidents should be risk-ranked as to their potential severity and probability of recurrence. Those with high potential probability and severity should be investigated.

Although an accident investigation is an after-the-event activity, it is vital in order to set up pre-contact controls to prevent the recurrence of similar accidents in the future. It analyzes the event, determines what went wrong, and paves the way for pre-contact controls to be implemented.

ACCIDENT RATIO

An often-ignored fact in safety philosophy is that more near-miss incidents occur than accidents. Some put the figure as high as 500 near-miss incidents to each serious injury. Also, more minor injuries occur for every serious injury. Another fact is that more property damaging accidents occur than injury resulting accidents.

The warnings provided by near-miss incidents, property damage accidents, and minor injuries should therefore not be ignored as they are indicators of larger things to come (Figure 2.1).

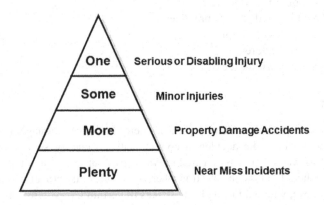

FIGURE 2.1   The updated accident ratio. [From McKinnon, Ron C. 2012. *Safety Management, Near Miss Identification, Recognition and Investigation.* Model 2.4. Boca Raton: Taylor and Francis. With permission.]

NEAR-MISS INCIDENTS

Regrettably, it is only the serious injury that is often considered worthy of investigation, yet the same event that causes minor injury may have had the potential to cause serious injury. The high-potential near-miss incident *could* have caused serious injury under slightly different circumstances. Accidents that result in damage also have potential to cause injury.

This makes them eligible for a complete investigation as in the case of a serious injury accident. Often the different outcomes of similar events are fortuitous. As McKinnon (2000) says

> The type of injury that results from an accident is largely dependent on factors that can neither be predicted nor controlled, and which numerous safety authors refer to as fortune or chance. H.W. Heinrich's (1959) one-line statement in *Industrial Accident Prevention* seems to have been grossly overlooked by the safety profession. They have ignored what he has said and based all efforts, recognitions, punitive measures, successes, and failures on the degree of injury as a result of an accident. Numerous studies have also concluded that there is little, if any, correlation between good safety controls, a good safety program, and the number and severity of injuries. Using the degree of injury as a measurement of safety performance or safety failure is completely misleading. The difference between minor injuries, disabling injuries, lost-time, and fatal injuries is largely a matter of luck.

(p.177)

INTERRUPTIONS

Adverse events such as accidents disrupt the business in one way or another. Workers stand around after an accident and ponder the situation or render assistance to injured workers. The site needs clearing and sometimes machines or equipment need repairs. Legal visits and accident investigation inspections also cause interruptions. Internal investigations and witness interviews disrupt the normal work day. In some serious injury or fatal accident cases, the worksite is shut down until all investigations are complete. These business interruptions cost money as production time is lost.

LEGAL REQUIREMENTS

It is a legal requirement in most countries to report all serious accident events, as well as some critical machine failures, to the local safety and health regulators. An internal accident investigation does not relieve the organization of these responsibilities. However, internal investigation of the serious injuries alone focuses on the tip of the iceberg, whereas the underlying problems and potential accidents lie in the base of the iceberg, below the waterline.

LESSONS LEARNED

After an accident, everyone has twenty-twenty vision. The circumstances leading up to the accident become clear and obvious. Good health and safety management systems should have built-in procedures and checklists to have twenty-twenty vision before the occurrence of the event. This will help the organization not to have to learn from painful and costly events. Accident prevention is the best proactive approach to workplace health and safety.

3 Accidental Loss Causation Theories

ACCIDENT CAUSATION

This question has been asked many times and each time the answers differ. Depending on which side of the fence the party is on, a finger is always pointed at the other party. Accidents open up a can of worms, and parties are quick to defend and eager to blame. Neither solves the problem. An accident is a problem. It needs to be solved, and finger pointing and blaming will not solve the issue. Accidents are caused and can therefore be prevented.

TRADITIONAL COP-OUTS

Traditional management had a number of cop-outs to absolve themselves from the safety liability issue. They were as follows:

CONTRIBUTORY NEGLIGENCE

Contributory negligence is a defense based on negligence. It applies where a plaintiff, through his or her own negligence, contributed to the harm he or she suffered. Should an employee not be wearing required personal protective equipment, and be injured as a result of this, this would be interpreted that he or she contributed to the injury.

ASSUMPTION OF RISK

The assumption of risk prevented the injured employee from claiming against the employer if the employer could prove that the employee voluntarily, and knowingly, assumed the risks at issue, inherent to the dangerous activity in which he or she was participating at the time of the injury.

MASTER AND SERVANT ACTS

During the 18th and 19th centuries, these acts were designed to regulate the relationship between employers and employees. The acts are generally regarded as heavily biased toward employers. The law required the obedience and loyalty from servants to their contracted employer, with infringements of the contract punishable before a court of law, often with a jail sentence of hard labor. This law made it almost impossible for an injured employee to claim any form of compensation from the employer for injuries sustained on the job.

ACCIDENTS ARE COMPLEX EVENTS

The International Labor Organization (ILO), *Encyclopedia of Occupational Health and Safety* (2020a), explains

> Accidents are defined as unplanned occurrences which result in injuries, fatalities, loss of production or damage to property and assets. Preventing accidents is extremely difficult in the absence of an understanding of the causes of accidents. Many attempts have been made to develop a prediction theory of accident causation, but so far none has been universally accepted. Researchers from different fields of science and engineering have been trying to develop a theory of accident causation which will help to identify, isolate and ultimately remove the factors that contribute to or cause accidents.
>
> *(ILO Website, Copyright © International Labor Organization)*

NO SINGLE CAUSE

If one reviews accidents that have occurred over the last couple of centuries, it becomes obvious that many factors contributed to the event. Seldom is an accident as a result of a single cause. Most accidents have multiple causes, which must be unearthed by effective accident investigation methods.

DOMINO ACCIDENT SEQUENCE

The occurrence of a preventable injury is the culmination of a series of events or circumstances, which invariably occur in a fixed and logical order. One is dependent on another, and one follows because of another. This constitutes a sequence that may be compared with a row of dominoes placed on end and in alignment with one another so that the fall of the first domino precipitates the fall of the entire row.

ACCIDENT CAUSATION THEORIES

Before being able to investigate an accident, one needs to know what an accident is, and how it is triggered, and be able to distinguish the different components. There are a number of schools of thought concerning the ideal model depicting the sequence of occurrences and circumstances that constitute an accident. Many are complex and many contain similar factors. Several theories on accident causation are discussed here to enable the investigator to establish an idea of causation factors and to understand some common characteristics that they possess.

HEINRICH'S DOMINO ACCIDENT SEQUENCE

The accident domino sequence was first proposed by H.W. Heinrich. In his book, *Industrial Accident Prevention (4th Edition)*, H.W. Heinrich (1959) first devised the theory that the accident sequence had five steps, and compared them to a row of falling dominoes (Figure 3.1).

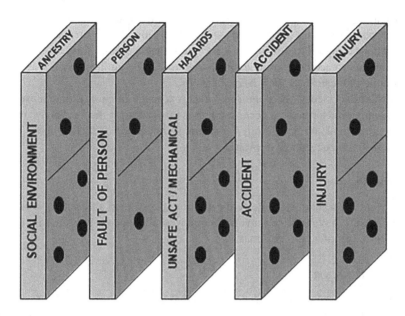

FIGURE 3.1 The Heinrich Domino Sequence. The term "accident" actually refers to the impact and exchange of energy that causes the loss.

Heinrich's theory was that the ancestry and social environment of the person caused recklessness, stubbornness, and other undesirable traits that may have been passed on through inheritance. This led to the fault of the person, the unsafe act, the unsafe condition, and the injury.

- He also proposed that the environment may influence undesirable character traits or may even interfere with the person's education. This was the *first* domino that triggered off the accident sequence.
- His *second* domino, the fault of person, referred to the inherited or acquired faults of the person such as a violent temper, nervousness, excitability, inconsiderate, and ignorance of safety practices. These, he stated, constitute the basic reasons for committing unsafe acts or for the existence, or creation of mechanical or physical hazards.
- Domino number *three* represented the unsafe act and/or mechanical and physical hazard. These were the unsafe performance of people, such as standing under suspended loads, horseplay, and rendering safety devices inoperable. Mechanical and physical hazards such as unguarded machines, the absence of handrails, and insufficient lighting or ventilation were also identified as the direct (or immediate) causes of the accident.
- The *fourth* domino, incorrectly termed the accident domino, is the *contact* and exchange of energy, such as the fall of a person, striking against, or other contacts with substances greater than the body threshold resistance.
- The end result of an accident, according to Heinrich, is the injury (*fifth* domino), which included fractures, lacerations, occupational disease, or fatalities.

Heinrich also proposed that 88% of all accidents were caused by the fault of the person, 10% were as a result of the unsafe environment, and 2% he attributed to acts of nature. This axiom led to a major misconception that accidents were caused by the worker and led to accident investigations seeking out the guilty party and ignoring the systemic root causes of the adverse event (p.21).

In his domino theory, the point of contact, exchange of energy, and subsequent injury or damage is termed the *accident*. This is misleading as the *total event*, including the contact and subsequent injury or damage is the *accident*. The injury is one phase of the total occurrence or event.

BLAME THE PERSON

If one accepts this antiquated accident causation theory, then all accident investigations will result in blaming the involved employee. This means that the root causes will never be identified nor eliminated.

NOT ONLY BEHAVIOR

What has further exasperated the situation is that the unsafe mechanical or physical conditions are often omitted from the immediate causes in this loss causation model (conveniently perhaps?) thus incorrectly laying full emphasis on the unsafe acts of workers.

Heinrich's recommendation of eliminating the unsafe act or the unsafe mechanical or physical environment to stop the domino effect was, however, correct.

FRANK E. BIRD'S UPDATED DOMINO ACCIDENT SEQUENCE

In 1992, Frank E. Bird, Jr., modified the domino accident sequence. He indicated that the reason for root causes existing was the lack of managerial control. This lack of control was in the form of an inadequate health and safety program, inadequate program standards, and/or failure to comply with those standards.

Bird also grouped the root causes of accidents into personal and job factors. He indicated that personal injury wasn't the only end result of an accident, but could include other losses such as property and material damage, loss to process, and/or damage to the environment.

NOSA'S UPDATED DOMINO ACCIDENT SEQUENCE

As indicated in the publication *Advanced Questions and Answers on Occupational Safety and Health*, the National Occupational Safety Association (NOSA) (1993) upgraded the domino sequence and added the stabilizer, which was the NOSA Management by Objectives (MBO) Five Star Safety and Health Management System. This system contained 73 crucial safety management system (SMS) elements with their standards. These standards, if introduced, would give management control over the *lack of control* domino, and therefore stabilize the personal and job factors, which in turn would eliminate the high-risk acts and high-risk work environment (pp.3–8).

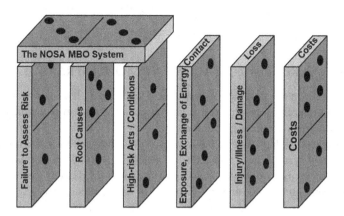

FIGURE 3.2 The NOSA Domino Sequence showing how the health and safety management system (The NOSA MBO System) stabilizes the dominoes, preventing the accident sequence from being triggered.

The safety standards, procedures, mechanisms, and processes to reduce the workplace risk on an ongoing basis are within the SMS. This is the stabilizing domino bridging and supporting the lack of control and basic and immediate causes, thus stabilizing the dominoes and preventing the sequence from happening (Figure 3.2).

A *sixth* domino was added in the form of costs. Past experience showed that if safety advisors and consultants could indicate to management that there were cost implications with every accident, they would be more willing to participate and buy into the SMS.

THE HUMAN FACTORS THEORY

The human factors theory of accident causation attributes accidents to a chain of events ultimately caused by human error. It consists of three broad factors that lead to human error:

- Overload
- Inappropriate response
- Inappropriate activities

THE EPIDEMIOLOGICAL THEORY

The epidemiological theory of accident causation considers industrial hygiene that concerns environmental issues and may result in sickness, disease, and other forms of health impairment. Epidemiology is the study of the relationship between the environment and diseases and can be used to study causal relationships between environmental factors and accidents.

THE SYSTEMS THEORY

The systems theory of accident causation is based upon the belief that an accident may occur as a system made up of a person or host, a machine or agency, or the environment. The likelihood of whether an accident occurs is determined by how the components of the system interact with one another.

FUNCTIONAL RESONANCE ACCIDENT MODEL (FRAM)

FRAM is based on the complex systemic accident theory. It considers that system variances and tolerances result in an accident when the system is unable to tolerate such variances in its normal operating mode.

According to Erik Hollnagel (2004), safety system variance is recognized as normal within most systems and represents the necessary variable performance needed for complex systems to operate, including limitations of design, imperfections of technology, work conditions, and combinations of inputs which generally allowed the system to work. Humans and the social systems in which they work also represent variability in the system, with particular emphasis on the human having to adjust and manage demands on time and efficiency (Website 2021).

THE COMBINATION THEORY

The combination theory of accident causation proposes that no one model or theory can explain all accidents. Factors from two or more causation models might be part of the cause.

MULTIPLE CAUSATION THEORY

The multiple causation theory is similar to the combination theory. It postulates that for a single accident there may be many contributory factors, direct causes, and sub-causes, and that certain combinations of these give rise to accidents. According to this theory, the contributory factors can be grouped into the following two categories:

- *Behavioral*
    This category includes factors relating to the worker, such as improper attitude, lack of knowledge, lack of skills, and inadequate physical and mental condition.
- *Environmental*
    This category includes improper guarding of other hazardous work elements and degradation of equipment through use, and unsafe procedures. The major contribution of this theory is to bring out the fact that rarely, if ever, is an accident the result of a single cause or act.

SINGLE EVENT THEORY

This theory proposes that an accident is thought to be the result of a single, one-time easily identifiable, unusual, unexpected occurrence that results in injury or illness. Some still believe this explanation to be adequate. It's convenient to simply blame the victim when an accident occurs. For example, if an employee is injured by a falling object, the worker's lack of attentiveness may be explained as the cause of the accident.

This simplistic theory gave rise to accident causes being listed as:

- Carelessness
- Lack of present moment thinking
- Failure to observe the obvious
- Failure to expect the unexpected
- Employee failed to follow procedures

Responsibility for the accident is placed squarely on the shoulders of the employees. This is a traditional approach based on archaic theories that the majority of accidents are caused by the unsafe behavior of the employee and leaves management free from responsibility. Accident investigations based on this theory will not identify and solve the real problems.

Ludwig Benner, Jr., (1978), says the following about the single event theory,

> This perception seems rooted in primitive history. If an unusual phenomenon occurred, and there was no ready explanation for it, the survivors sought a scapegoat as the "cause" of the occurrence. Find the "cause" (read scapegoat) and the victims are satisfied.

(p.1)

PETERSEN'S ACCIDENT/INCIDENT THEORY

This theory was developed by Dan Petersen and states that the causes of accidents/incidents are because of human error and/or system failure. Human error is due to overload, traps, and decision to err. Human error may directly cause the accident or may cause system failure, which may cause the accident resulting in injury or loss.

ENERGY RELEASE THEORY

Dr William Haddon, Jr., theory of accident causation and control, portrays accidents in terms of energy transference. This transfer of energy, in large amounts and/or at rapid rates, can adversely affect living and non-living objects, causing injury and damage.

The thesis is that accidents are caused by the transfer of energy with such force that bodily injury and property damage result. According to Dr Haddon, strategies can interrupt or suppress the chain of accident-causing events. These strategies revolve around control and prevention of:

- A build-up of energy that is inherently injurious
- The creation of an environment that is not conducive to injurious build-up of energy
- The production of counteractive measures to injurious build-up of energy

SWISS CHEESE MODEL

The *Swiss Cheese Model* of accident causation, originally proposed by James Reason (2016), likens human system defenses to a series of slices of randomly-holed Swiss cheese arranged vertically and parallel to each other with gaps in between each slice.

Any component of an organization is considered a slice of cheese.

- Management is a slice
- Allocation of resources is a slice
- An effective safety system is a slice
- Operational support is a slice

If there are any deficiencies or flaws in any of these slices of an organization, then the organization will have a hole in that slice. Reason was able to construct his integrated theory of accident causation through in-depth research into the nature of accidents, leading him to the following insights:

- Factors can range from unsafe individual acts to organizational errors.
- Accidents are often caused by the confluence of multiple factors.
- Many contributing factors to an accident are latent errors – they're lying dormant waiting to be triggered by any number of active errors.

Reason hypothesizes that most accidents can be traced to one or more of four levels of failure:

- Organizational influences
- Unsafe supervision
- Preconditions for unsafe acts
- The unsafe acts themselves (Website 2021)

THE SYMPTOMS VERSUS CAUSES THEORY

The symptoms versus causes theory is not so much a theory as a warning to be heeded if accident causation is to be understood. Usually, when investigating accidents, investigators tend to focus on the obvious causes of the accident and overlook the root causes. Unsafe acts and unsafe conditions are the symptoms (the proximate or immediate causes) and not the root causes of the accident.

THE PURE CHANCE THEORY

The pure chance theory states that everyone of any given set of workers has an equal chance of being involved in an accident. It further implies that there is no single discernible pattern of events that leads to an accident. In this theory, all accidents are treated as "beyond normal control" (almost as acts of Providence) and that there exist no interventions to prevent them.

CAUSE, EFFECT, AND CONTROL OF ACCIDENTAL LOSS

When proposing the *Cause, Effect, and Control of Accidental Loss* accident sequence, which is based on the Heinrich accident domino sequence, McKinnon (2000) stated,

An accident is initiated by failure to assess the risk of either behavior or work conditions. This triggers off basic causes in the form of personal and job factors, which then lead up to unsafe acts and unsafe conditions. These unsafe acts and unsafe conditions are the immediate cause of the event. Luck Factor 1 determines whether the unsafe act or unsafe condition will result in a contact with a source of energy or merely be a near miss incident. In this case, nothing happens but there was potential for loss. Luck Factor 2 then determines the outcome of the contact with a source of energy. The outcome could be injury, illness, or disease, damage to machinery, property, materials, vehicles or a business or process interruption. Luck Factor 3 then determines the severity of the injury, as once the exchange of energy has taken place, we have no control of the severity of the consequences.

(p.212)

Accidents occur when a series of small blunders come together at a certain time and under certain circumstances. Three luck factors determine the final outcome of the event once it has been triggered (Figure 3.3).

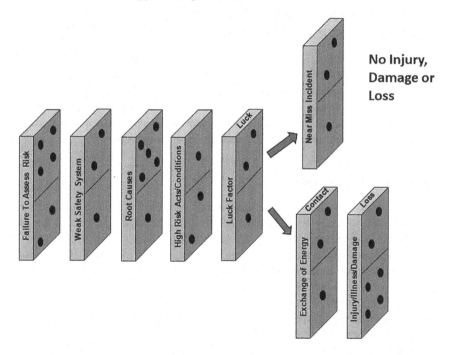

FIGURE 3.3   Once a high-risk behavior takes place, or high-risk or condition exists, Luck Factor 1 determines if the outcome is a near-miss incident or a loss in the form of injury or damage. (From McKinnon, Ron C. 2012. *Safety Management, Near Miss Identification, Recognition and Investigation*. Boca Raton: Taylor and Francis. With permission.)

ACCIDENT PRONENESS THEORY

According to the ILO, *Encyclopedia of Occupational Health and Safety* (2020a), the accident proneness theory maintains that within a given set of workers, there exists a subset of workers who are more liable to be involved in accidents. Researchers have not been able to prove this theory conclusively because most of the research work has been poorly conducted, and most of the findings are contradictory and inconclusive. This theory is not generally accepted. It is felt that if indeed this theory is supported by any empirical evidence at all, it probably accounts for only a very low proportion of accidents without any statistical significance (Chapter 56, Copyright © International Labor Organization 2020a).

PURPOSE OF AN ACCIDENT INVESTIGATION

The purpose of an accident investigation is to ascertain the facts, irrespective of how painful and exposing they are, and to delve into all the factors relating to the event to identify the root causes, and to propose solutions for them to prevent a recurrence of similar events in the future.

INVESTIGATION FOLLOWING THE LOGICAL SEQUENCE ACCIDENT INVESTIGATION METHOD

Effective accident investigations start at the end result of the event, with the losses. All losses are first determined. The impact or energy exposures are then identified. The proximate or immediate causes of the unplanned release of energy in the form of high-risk actions, unsafe work environment, or acts of nature are then ascertained. Working back from each immediate cause, a root cause analysis is done by asking *why* the act was committed or *why* the work environment was unsafe.

ROOT CAUSE ANALYSIS

Root cause analysis is a structured questioning process that enables investigators to recognize and discuss the underlying beliefs, practices, and processes that result in accidents. Root causes are basic accident causal factors, which if corrected or removed, will prevent the recurrence of an accident. The answers to these questions will deliver the root accident causes. The root causes will also indicate where the organization failed to identify the risks and put suitable safety systems in place, or maintain them, to prevent these root causes occurring.

In the event of a high-potential near-miss incident, the exact same investigation technique is followed except that as there was no loss, the potential loss is estimated. The flow of energy is determined as well as the potential exchange that could have taken place.

REMEDIAL MEASURES

Only once all the root causes have been identified, can failures and inadequacies in the system be determined and rectified. Some issues can be identified within a short time period, and others such as retraining may take longer.

SUMMARY

Over the years, many theories about how accidents are caused have been proposed, each with their own merit; however, one factor they have in common is that accidents are caused and don't just happen. Another common thread is that there is always more than one cause for an accident, and accident investigation must uncover all the causes, both direct and indirect. Although some theories focus heavily on the behavior of employees involved in the accident, effective investigation delves deeper to find out what spurned those behaviors at the time of the event.

4 Traditional Accident Investigation

WHO MESSED UP?

Traditional accident investigations normally revert to the obvious easy way-out question, "Who messed up here?" Traditional thinking and safety theories tend to lead to a witch-hunt after an accident occurs. The organization looks for a person to blame for the event so they can absolve themselves from any wrongdoing or blame.

ARCHAIC APPROACH

There is also an emotional reaction to an event that can sometimes be traumatic for fellow workers, and the organization itself. With this traditional and archaic approach to safety, blaming someone seems to take the pressure off the organization and solve the accident problem. Nothing could be further from the truth. During hundreds of accident investigation training workshops, I have asked the attendees to show me one person who has never "messed up," be it at home, in school, or at work. Everyone agrees that we all mess up on a frequent basis. Messing up is not the sole cause of an accident. It may be a contributing factor but is not the deep-seated root cause.

PUNITIVE ACTIONS

The words accident and discipline have become synonymous in safety. Discipline is thought to be the solution to workplace accidents. Some safety cultures require some form of discipline after every event. This does little to solve the problem and does more harm than good. It just drives the safety problems underground. It discourages accident reporting and lulls the organization into a false sense of security, thinking that they are accident-free. In fact, they are just not hearing about the accidents because of the fear of discipline.

DISCIPLINE

Writing in *Industrial Safety and Hygiene News*, Dave Johnson (2018), quotes Jim Frederick of the United Steelworkers who said

> Employing discipline as the automatic, indiscriminate default response to safety problems prevents employers from engaging with their workers to identify the real causal factors involved with the injury. As a result, the prevalence of under reporting of injuries remains. Controlling workplace health and safety hazards is exponentially increased when the hazards are unknown because injuries are not reported (or investigated) due to the chilling effects of management systems that lean hard on discipline.

(ISHN website 2021)

THE SAFETY FEAR FACTOR

The safety fear factor is real and exists in most workplaces around the globe. After interviewing numerous groups as to whether or not they would report an injury caused by an accident in which they were involved, the answers were always no. Further questioning revealed that they expressed a fear of losing their job should they have been involved in an accident. This fear is real and leads to hiding injuries, or claiming injuries as home, or off-the-job sport injuries. In many instances, the injured employee pays for their medical treatment out of their own pocket. This is the safety fear factor, and because of it, management seldom gets the full picture of what is happening at the workplace. If accidents are not reported, they cannot be investigated and the problems cannot be identified and solved.

FACT FINDING AND NOT FAULT-FINDING

It has been written over and over and stated repeatedly for years, and although many hear it, many do not practice the fact that accident investigation is a fact-finding mission and not a fault-finding exercise. Finding fault and singling out an individual, or individuals to blame, after an accident, does not identify the cause of the problem, nor does it offer a solution to it. If accident investigation is to be effective, it must find all the facts relating to the event. It should not get bogged down in finding who was at fault. Only by finding the facts and then tracing them back to the origins of the event, can the investigation be effective.

COVER-UPS

Because of the tendency to blame individuals for accidents and to find who was at fault, many accidents are covered up, and the facts and circumstances are hidden to protect those involved. Colleagues and peers will also assist in the cover-up to protect their friends. The safety fear factor is responsible for these cover-ups. Breaking the safety record and disrupting the day-to-day business because of an accident are all unwanted occurrences, and everything is done to avoid the shame of being singled out as the person who caused this. This makes effective accident investigation almost impossible.

COP-OUTS

An accident becomes a hot item that most want to divorce themselves from and not get involved in. This has led to a situation where employees are normally blamed if an accident occurs, property is damaged, or somebody is injured. Leadership frequently cops out when this happens. The roots of this situation go back to H.W. Heinrich's theories and findings. In his book, *Industrial Accident Prevention: A Scientific Approach* (1959), Heinrich stated that research that he had done, using the most prominent accident causes, showed that 88% of accidents were caused by unsafe acts of people (p.21).

BLAME FIXING

Accompanying accident cover-ups are accident cop-outs. After an accident, a finger pointing exercise happens. No one wants to accept responsibility for any part of the event. Management often blames workers for the accident, and workers blame the management. Seldom does supervision admit failure that led to the accident. Because of repercussions after the accident, management wants to distance themselves from the event and often tries to put the blame entirely on the injured employee.

Heinrich's statement (first made in 1931) that unsafe acts are responsible for the majority of accidents is largely to blame for this dilemma.

Ashley Johnson (2011), writing in *Safety and Health* magazine, discusses Fred A. Manuele's analysis of Heinrich's axioms and wrote

> Another issue Manuele raised was that the original files Heinrich used do not exist, preventing others from reviewing his work. In refuting Heinrich's work, Manuele singled out the 88-10-2 ratio of accident causation as having the most influence and causing the most harm to the safety profession.
>
> "Why harm? Because when basing safety efforts on the premise that man failure causes the most accidents, the preventive efforts are directed at the worker rather than on the operating system in which the work is done."

(Website 2021)

WHY ACCIDENTS AND NEAR-MISS INCIDENTS ARE NOT REPORTED

The answer to this is simple. If the organization believes that the employee is the cause of the majority of accidents, then the employee must be dealt with. Employees have a fear of losing their jobs and this fear is real, even though health and safety legislation seemingly protects employees from this. Employees that I have interviewed have given the following reasons why they would not report an accidental injury:

- Loss of "safety" bonus
- The fear of punitive action
- Terrified of being the one to break the safety record
- Possibility of losing their job
- Fear of being overlooked for promotion
- Embarrassment

SAFETY BONUS

One of the biggest handicaps to the reporting of accidents is the safety bonus. This is normally a cash payout for "working safe" over a certain prescribed basis. What it really is, is a bonus for *not being injured* over a certain period. It is based on downstream measures of injury and is thus not a safety bonus. Injured employees would rather hide their injury (or take it home) than lose this bonus.

Writing in *EHS Today*, Philip La Duke (2011), listed fear as one of the nine reasons near miss incidents are not reported:

> Fear – believe it or not, may actually be the least common reason workers avoid reporting near misses. It's true that some workplaces cultivate an environment where employees are punished for being injured, so these workers are unlikely to report near misses if they fear they will lose their jobs. Overall, however, this usually isn't the most common reason workers neglect to report their near misses.

(Website 2011)

THE SAFETY RECORD

The number of days worked without injury is the archaic method used by many firms to measure their degree of safety. The workplace may be full of hazards, but as long as there is no injury the organization accepts that it is safe. As the cumulative injury-free hours mount, so does the anxiety of the employees. Who would like to be the one who spoils the safety record? This is a definite deterrent to the reporting of the accident and leaves no incentive whatsoever for reporting other events such as property damage accidents and near-miss incidents (Figure 4.1).

FIGURE 4.1   Some of the main reasons for the non-reporting of workplace accidents.

HISTORIC RESTRAINTS

The major downfall of accident investigation lies perhaps in the historical carry-over paradigm that the injured worker is the one who caused the accident, and where the accident fault lies. Employees have been punished for being involved in accidents. Some have been sidelined and some have been branded as accident-prone employees. This historic belief of blaming the worker has taken the focus away from fixing the unsafe workplace to rather trying to "fix" the unsafe worker. This is impossible, and discipline, finger pointing, and fault-finding will not help prevent accidents.

PARADIGMS

The belief that "accidents just happen" is a paradigm. That the unsafe behavior of employees is responsible for accidents is too simplistic, and is also a paradigm.

The more we practice these paradigms, the further we get from the truth.

These paradigms need to be shifted before accident investigation can be meaningful and effective, and can determine all the causes of the event. This is to ensure that recurrences can be prevented.

SAFETY CULTURE

The safety culture of an organization should be one where there is no shame apportioned to being involved in an accident. A climate should exist where workers can feel comfortable to report accidents without the fear of repercussions. Fear, threats, and belittlement should not follow an accident. Rather a positive approach and a "how can we solve this problem?" attitude should prevail.

SUMMARY

The traditional approach to accident investigation was based on the myth that the majority of accidents were the fault of the worker. This was a convenient belief as it left management free from responsibility. What resulted was that all efforts were directed at the worker and the real causes of the accident were left uncovered and unrectified.

Safety bonuses and safety records create a climate that discourages the reporting of accidents, as workers are afraid of being the culprit who was responsible for breaking the safety record, or loss of the safety bonus. Punishment and ridicule following accidents have also led to non-reporting. All these safety misconceptions have created the safety fear factor among employees.

If accidents and other adverse events are not reported, they cannot be investigated and the underlying problems cannot be identified nor rectified.

5 The Politics of Accident Investigations

An accident investigation is a sensitive issue. Employees may have been injured, property may have been damaged, and an interruption to the day-to-day processes may have occurred. Legal entities may have to be involved, and sometimes bad press coverage could follow after the event.

STUMBLING BLOCK

Internal company politics is the biggest stumbling block in accident investigations. Management does not want to be found responsible for the accident, and employees involved do not want to be blamed for what happened. So immediately there is a stand-off, hovering around the outcomes of the investigation.

OPENING A CAN OF WORMS

Few, if any guidelines on an accident investigation warn the investigators that they will be opening a can of worms when they investigate an accident. Investigators be aware! Opening a can of worms means to create a complicated situation while doing something to correct a problem. While seeking the source of the problem, it leads to many more problems. While investigating root accident causes, many more underlying causes may surface which may indicate a deeper underlying problem. Sometimes, the uncovering of these deep-seated issues is not welcomed by the organization.

BLAME GAME

Traditionally, accident investigations were carried out to fix blame and execute punitive measures on the injured and involved workers. Many safety cultures still believe in blaming individuals who happen to be injured in workplace accidents. This blame invariably led to some forms of discipline. This archaic approach still exists in many workplaces, and although often denied, is still an underlying trend in many instances. Once blame is fixed, the problem is seemingly solved, and it's back to work as usual. Blame fixing does not identify the real causes and neither does it fix the problems.

Wilson Bateman (2018) says

> Some people consider a close call to be a free lesson and a chance to look at the safety management system. This is not always the case. Sometimes, a close call or incident is followed by "The 2nd Incident," or as I call it, the "Name, Blame and Shame Game." In other words, a quick fix, throwing someone under the bus, safety

DOI: 10.1201/9781003220091-6

cover-up, white wash, a conspiracy of sorts, usually with good intentions. In this type of incident investigation, someone must be blamed because corporate wants a resolution. It's as old as time itself, but rarely is it ever effective.

(website 2021)

LOOKING GOOD

Organizations want to look good. They want to be seen as institutions that care for employees. An accident with resulting injuries is a devastating occurrence and makes them look bad. Reputations are on the line. Shifting the blame for the accident onto the employees makes the employees look bad, but the company saves face and looks good. This political ploy hampers effective and meaningful investigations.

MEASURES OF A SAFE WORKPLACE

Many workplaces are stuck in the paradigm that safety is a measure of the number of injuries resulting from work-related accidents. The result of this is that an injury reflects directly on the safety of the workplace. The safety "record" is lost and safety falls in the spotlight. The more injuries that occur, the more unsafe the workplace appears. This is a strong incentive to shift blame and absolve the company from the accident. They want to distance themselves from the accident causes and shift the blame to employees. This drives accidents underground. They are not reported nor investigated, so the organization thinks it's doing well.

SAFETY RECORDS

The safety record is a source of pride and joy in workplaces which they regard as the sole measure of their safety success. Any accident with resultant injury is a threat to this record, and reactions to the person, or persons who ruined the record are inevitably negative. I have often asked the question to groups of employees, "Who here would like to be the person who is injured and breaks the company record of working 1.5 million workhours without injury?" Not surprisingly, no one raises their hand!

SAFETY RECORD DISPLAY BOARDS

Safety record boards displayed around the workplace indicate how many days the organization has worked without a serious injury. These constant reminders make it clear that the injury-free record is important. This adds to the pressure on employees not to be injured. No one would like to be the one who is injured and spoils the record.

PEER PRESSURE

Peer pressure is a strong activism factor in the workplace. If all share a cash payout in the form of an injury-free reward, then the closer the organization moves to its injury-free goal, the more the pressure. If an employee is injured, all lose the safety bonus, so there is immense pressure exerted by the group to ensure that no one is injured (or no one *reports* a workplace injury) before the bonus is paid out.

LEGAL ACTION

In most countries, occupational health and safety law requires the reporting of serious injury accidents and in some cases reportable machinery failures or releases. Once reported, this may result in an on-site investigation by these legal agencies, and depending on the circumstances, may result in heavy penalties for safety violations that led to the accident. This hampers internal accident investigations, as in some instances, the organization is afraid that the legal agencies may use the organization's internal investigation findings against them.

SHIFTING THE RESPONSIBILITY

Management cannot and should not shift the responsibility for accidents from them to the employees. Irrespective of what went wrong, who did what, or who didn't do what, management is ultimately responsible for the safety at the workplace and must take the brunt of all accidents.

Often the injured worker is only a pawn in the safety management system. Most of the time he or she is an innocent victim. Placing blame for accidents is really shifting the responsibility from where it belongs, with senior management.

Although many may have erred and contributed to the causes of the accident, the organization should ask, "What more can we do to prevent a recurrence of this event?" Shifting responsibility and seeking a way out will not solve the problem, nor will the politics of shifting blame.

RESULTS OF CONFLICTING INTERESTS

After an accident, a weak or immature safety culture results in conflicting interests. The results of these conflicting interests and separate agendas are that the accident facts are sometimes distorted and evidence is modified or changed to suit one particular agenda. In fact, a cover-up occurs. This is intended to shift responsibility for the accident to a different party. This makes it difficult for investigators who do not get the true picture of how the accident occurred, and what root causes need to be rectified. Organizational politics often come into play after an accident, especially if there is not a positive, proactive safety culture present in the organization (Figure 5.1).

FIXING THE PROBLEM, NOT THE PERSON

The objective of an accident investigation is to examine the event to uncover the direct and indirect (root) causes, and propose measures to rectify these deficiencies. This will contribute to fixing the problem. It is not intended to fix the worker by allocating blame and instituting disciplinary measures. This will not solve the problem. Investigations should be unbiased and focus on the real causes, the underlying human and work environment root causes.

FIGURE 5.1 A weak safety culture contains separate and opposing agendas.

CONCLUSION

An accident investigation is a sensitive issue. Employees may have been injured, property may have been damaged, and an interruption to the day-to-day processes may have occurred. Legal entities may have to be involved and sometimes bad press coverage could follow after the event.

An accident can have an upsetting influence on all within an organization as nothing can be done to undo the harm done. Employees become emotional and the fear factor prevails as they are afraid of repercussions after the accident. An accident investigation is aimed at uncovering the true facts of the event, and this may open a can of worms.

A cover-up and distortion of facts, and even contamination or readjustment of the accident site may occur. Because employees are afraid, they may not be truthful during the witness interviews, and management, not wanting to be found to be at fault, may do the same.

Section II

Accident Investigation Methodology

6 Objective of Accident Investigation

A SYSTEMATIC REVIEW

The objective of both accident and near-miss incident investigation is to carry out an investigation into the undesired event to determine what happened and what can be done to prevent a similar event recurring.

In a nutshell, an accident investigation is a systematic review and analysis of the event, with the aim being to identify and pose solutions to the problem. An accident investigation is carried out to establish where the system failed rather than where the worker failed.

If positive preventative measures are not taken after an accident or high-potential near-miss incident, the probability of a recurrence is great. The investigation procedure would lead to the immediate and root causes of the event and help determine what steps should be taken to prevent a similar accident or near-miss incident.

LEGAL REQUIREMENT

Most occupational health and safety legislation around the world requires that accidents involving serious injury or fatality be investigated and that actions be taken to prevent the occurrence from happening again. Many of these events are also required to be reported to the local regulators. An objective of an accident investigation, therefore, is to fulfill an organization's legal obligation to investigate accidents at the workplace. The investigation also indicates to employees and the general public that the organization shows concern about the accident and this helps relationships between employees and management.

SAFETY CULTURE

A positive safety culture is one in which accidents are accepted as (unfortunate) opportunities to identify and solve problems which could be worse the second time around. The culture would engender an integrated Safety Management System (SMS), frequent incident and accident recall sessions, and formal training on an accident investigation. The safety culture should welcome the findings of the investigation and be diligent in putting control measures, recommended by the investigation, to prevent similar events from happening again, into action. A positive safety culture would view the investigation as fact-finding and not fault-finding.

DOI: 10.1201/9781003220091-8

WHAT OCCURRED?

Accident investigations establish *what* happened, *where* it happened, *what* caused it to happen, and *what* the consequential losses were? It paints a complete picture of the accident and reveals all the facts backed up by tangible evidence. The investigation then poses remedial measures to be taken to prevent a similar event happening in the future.

EVENTS

The investigation is designed to uncover all the events leading up to the accident and to identify the immediate and root causes of these. Since the whole purpose of the reporting, investigation, analysis, and evaluation of an accident is to prevent and control the recurrence of similar accidents, near-miss incidents, injuries, property damage, and other losses; all events must be scrutinized during the investigation.

TIMELINE

Another objective of an accident investigation is to establish a timeline. A time-line is a list of significant events arranged in the sequence in which they happened. Timelines explain what occurred during a certain period of time, starting with the earliest, (initiating) event, and moving forward through time to the conclusion of the event. These events, no matter how trivial, occur on a certain timeline which must be established by the investigator.

EVIDENCE

Another objective of prime importance in an investigation is the gathering of evidence. There are many forms of accident evidence and a few are discussed below.

ACCIDENT SITE

The prime source of information about the accident comes from the physical site(s) of the accident. This may even be more than one physical location. A record must also be made of the environmental conditions prevailing at the time of the accident.

DOCUMENTED EVIDENCE

Documented evidence appertaining to the accident is vital as it will help fill in the gaps and answer questions raised by the investigators. The objective of an investigation is to obtain and examine as much documented evidence relating to the accident as possible.

WITNESSES

The investigation should also identify all possible witnesses and ensure that they are interviewed about everything in relation to the event.

SYSTEM FAILURE

An accident is a result of a system or systems failure. An investigation is carried out to uncover this failure. The failure could be a weakness in the hazard identification and risk assessments (HIRAs), which had been done or not done in the past. One or more of the SMS elements could prove to be missing or inadequate or have failed, which allowed the event to occur. The advantage of an accident investigation is it determines whether there were adequate or inadequate health and safety system standards and procedures in place, where they failed, or why they were not followed. One of the most beneficial objectives of an accident investigation is that it contributes to the improvement of the SMS elements and standards.

WORKER FAILURE

Following an accident, many traditional investigations revert to finding out where the worker failed. This is not the aim of an accident investigation. In most accident scenarios, it will be found that people committed high-risk acts, omitted to follow procedures, took shortcuts, or in one way or another, behaved in an unsafe manner. This is not surprising as workers are human beings and do take chances, bypass systems, and take risks.

GETTING THE JOB DONE

When interviewed after an accident where a number of employees were injured, it was revealed that the employees were intent on "getting the job done." There was seemingly such pressure on getting the job done, that safety measures and rules were ignored in favor of getting the job done.

The objective of the investigation is to identify and record these high-risk behaviors and acts but not to judge them. Once all the facts have been established, an immediate and root cause analysis can begin.

FIXING THE WORKPLACE, NOT THE WORKER!

The objective of an accident investigation is not to fix the worker. In his paper entitled, *Fixing the Workplace, not the Worker*, Bill Hoyle (2005) writes,

> When you read a newspaper account of an industrial accident it will almost always conclude that the cause of the accident was worker error. In a society largely based on individualism, the idea that worker mistakes are the primary cause of accidents rings true with most people. There is no denying that workers make mistakes. However, in every industrial accident there are almost always several management safety systems involved which may not be readily apparent.
>
> Our safety management program centers on fixing the workplace, not the worker. This is done through creating effective safety systems. A strong process safety management system is one part of a systems approach to health and safety. Examples of safety systems include the safe design of processes and equipment, proper maintenance and inspection, and having effective procedures and training programs.

(p.3)

FIX THE PROBLEM

Accidents are investigated so that the causes of the problem are identified, examined, and solutions put in place to prevent a recurrence of the same or similar event in the future. Punishing employees and taking punitive actions against those involved will not solve the problem. One of the safety paradigms is that there must be discipline after every accident. Discipline will not solve the problem and will drive the reporting of future accidents underground.

According to the Canadian Centre for Occupational Health and Safety (CCOHS) (2019a) *Incident Investigation* factsheet:

> A difficulty that has bothered many investigators is the idea that one does not want to lay blame. However, when a thorough worksite investigation reveals that some person or persons among management, supervisors, and the workers were apparently at fault, then this fact should be pointed out. The intention here is to remedy the situation, not to discipline an individual. However, never make recommendations about disciplining anyone who may be at fault. Any disciplinary steps should be done within the normal personnel procedures.

> *(CCOHS Website 2020)*

FOLLOWING UP

Of prime importance is the follow-up after the accident investigation to ensure that the recommendations of the investigation have been implemented and are being maintained. The end objective of the investigation is to close the loop and ensure the weaknesses discovered that led to the accident have been rectified. Since the whole aim of an accident investigation is to fix the problem and make the workplace safer, the final question is, "Have the causes of the problem been identified and fixed, and will this prevent a recurrence of the same accident?"

CONCLUSION

The objective of an accident investigation is to carry out an investigation into the undesired event to determine what happened and what can be done to prevent a similar event recurring. It is a systematic review and analysis of the event, with the aim being to identify and pose solutions to the problem.

Accident investigations establish *what* happened, *where* it happened, *what* caused it to happen, and *what* the consequential losses were. It paints a complete picture of the accident and reveals all the facts backed up by tangible evidence. The investigation then poses remedial measures to be taken to prevent a similar event happening in the future. One most beneficial objective of an accident investigation is that it contributes to the improvement of the SMS.

7 Nominated Accident Investigators

The supervisor or manager directly in charge of the area in which the accident occurred should be the prime accident investigator. Delegating this responsibility to the safety department or safety coordinator is not acceptable. The safety department should only assist management in investigating the accident, as they have experience in this field and can be used as a resource.

MANAGEMENT'S RESPONSIBILITY

Management hires employees. They provide the necessary training and supervision, approve promotions and salary increases for the individual worker, and issue the work assignments, so, when an employee is injured, it is obvious that management should investigate what went wrong, *not* the safety department. Handing the task of investigation of an accident to the safety department is abdicating responsibility.

A JOINT APPROACH

Management may nominate and appoint an accident investigator within their areas of responsibility. It is essential that both management and the workforce must be involved in the investigation. A joint approach reinforces the message that the investigation is for the benefit of everyone.

MANAGEMENT TRAINING

Many ask why managers should investigate accidents within their areas of responsibility, and argue that managers have not been trained in accident investigations. Why not? Managers at all levels should be trained in safety management and accident investigation as well. This will equip them with the basic skill needed to investigate accidents. The safety department can always be used as a resource to assist in the investigation.

EXTERNAL EXPERTS

Depending on the circumstances, a subject matter expert may be called in to assist the investigator. This could be an expert in the techniques of accident investigation or an expert in the specific field relating to the accident.

HEALTH AND SAFETY REPRESENTATIVES

In some cases, the Health and Safety Representative may be appointed as the accident investigator. In fact, Health and Safety Representatives should be party to all investigations. While it is always beneficial to have worker representation on an investigation, this should not free the immediate manager of his or her responsibilities to investigate the accident.

POTENTIAL FOR LOSS

The extent of the resultant loss should not determine who should carry out the investigation. Past experience has shown that once there are fatalities, multiple fatalities, or great financial loss as a result of an accident, only then do top management get involved. What should determine participation in the accident investigation procedure is the *potential* for loss of the event *or what could have happened* under slightly different circumstances.

The difference between a near-miss incident with no loss, and an accident with great loss, is only a matter of luck. The near-miss event should receive as much attention as an accident that resulted in severe injury.

INVESTIGATION COMMITTEE

Depending on the circumstances, it might be beneficial to appoint a committee or sub-committee to investigate certain accidents. These committees could use problem-solving techniques to identify the root causes of the accident and propose remedial measures and actions to take to prevent recurrence. Committees established for this purpose should not be too big, and they should be given a definite timeline within which to complete the investigation. The committee could comprise:

- Employees with knowledge of the work
- The supervisor of the area, work, or person
- The safety coordinator
- Members of the Health and Safety Committee
- The Health and Safety Representative
- Worker union representative
- Employees with experience in investigations
- Representative from local government or police

EXPERIENCE

The investigating party or parties should have some experience in the process or procedures within the accident area. Ideally, they should have some training in accident investigation and should be aware of the organization's accident investigation standards, policies, and protocols. Preferably, they should have knowledge of accident causation models, and have some investigative experience.

KNOWLEDGE

The nominated investigator should have knowledge of occupational health and safety fundamentals, be familiar with the work processes, procedures, and skills employed in the area. They should also know the legal requirements applicable to the specific industry, mine or process, and be able to analyze the data gathered to determine causes, and propose recommendations. Investigators should have the skills to use interview, and other person-to-person techniques effectively, such as conflict resolution and questioning techniques, when they interview witnesses.

MANAGEMENT SKILLS

Accident investigators also require management skills based on the four main functions of management: *Planning, Organizing, Leading, and Control.*

- Was there a deficiency in the planning process that contributed to the problem? (*Planning*).
- Were there wrong, or under-skilled employees, in positions that created the problem? (*Organizing*).
- Was inadequate or incorrect leadership a contributing factor? (*Leading*).
- What safety management system (SMS) control systems were missing, inadequate, failed, or bypassed at the time the event occurred? (*Control*).

GUIDELINES

The organization should have guidelines and standards relating to accident prevention and accident investigation, these should include

- The written SMS standard pertaining to an accident investigation
- Training program syllabus for an accident investigation
- Initial accident report form and protocol
- Accident investigation form

INVESTIGATORS MUST BE COMPETENT!

Investigating an accident calls for a variety of skills and knowledge. It calls for dedication and a sense of urgency to get to the root causes of the problem. A nominated accident investigator, or team of investigators, should possess sufficient knowledge and experience to be able to follow the logical accident investigation sequence and uncover the root causes of the event and, in turn, recommend effective remedial measures.

CONCLUSION

The supervisor or manager directly in charge of the area in which the accident occurred should be the prime accident investigator. Management hires employees. They provide the necessary training and supervision, they approve promotions and salary increases for the individual worker, they issue the work assignments, so when an employee is injured, it is obvious that they should investigate what went wrong.

The nominated investigator should have knowledge of occupational health and safety fundamentals, be familiar with the work processes, procedures, and skills employed in the area. The safety department should be used as a resource to assist and advise. In some cases, the Health and Safety Representative may be appointed as the accident investigator. It is essential that both management and the workforce must be involved in the investigation. A joint approach reinforces the message that the investigation is for the benefit of everyone. Depending on the circumstances, it might be beneficial to appoint a committee or sub-committee to investigate certain accidents.

8 Problem-solving Methods

An accident investigation is a problem-solving technique. Accidents are the end result of some form of problem occurring, and the same methods used to solve workplace problems can be applied to solving accident problems. Accidental events are not part of the organization's planning process and are therefore deviations from normal practice. As with deviations from production processes, accident problems must also be solved. This is the purpose of an accident investigation.

There are a number of problem-solving techniques used in an accident investigation. It is for the investigators to decide which specific method they wish to follow. They may also decide to use a combination of methods.

BASIC PROBLEM-SOLVING STEPS

A basic problem-solving approach is as follows:

- Identify and clearly define the issue to be solved
- Gather the facts, then the immediate and root causes
- List the possible solutions
- Evaluate the options
- Select an option or options
- Document the agreed solutions
- Apply the solutions
- Follow-up

IDENTIFY AND CLEARLY DEFINE THE ISSUE TO BE SOLVED

The problem can only be solved once all the data relevant to the problem is obtained and arranged in a logical order (Figure 8.1). In an accident investigation, this would include information from the accident site, the witness information, and data obtained from the relevant documentation. The problem must be clearly defined and stated so that the investigation can focus on finding a solution.

GATHER THE FACTS, THEN THE IMMEDIATE AND ROOT CAUSES

Major deviations from prescribed norms and standards must be identified and noted. At this stage, the facts must be obtained and no opinions should be formed. The immediate causes of the event may seem obvious, but the main aim of the accident investigation is to delve into the causes behind the obvious causes and identify the root causes.

This step involves the accident site inspection, the witness interviews, and the relevant documentation scrutinized. All the high-risk behaviors and high-risk workplace

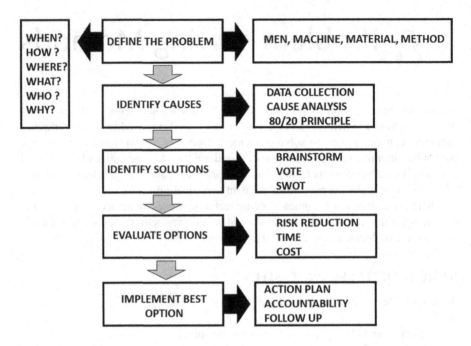

FIGURE 8.1   A basic problem-solving method.

conditions are recorded, and an immediate cause analysis is completed. Once this has been completed, the root cause analysis begins, and root causes are found for the high-risk behaviors, and then for the high-risk workplace conditions. Root causes have two main categories, those relating to the *personal factors* (human factors) and those relating to the *job factors* (organizational factors). Acts of nature are also a category of root causes.

LIST THE POSSIBLE SOLUTIONS

Only once all the facts of the event have been obtained, can the investigators start posing solutions to address the defined problems. Ideally, all involved in the accident investigation should be consulted on possible solutions to the problem. Both short-term and long-term solutions should be proposed at this stage. Brainstorming is a good way to gather solutions and alternate solutions, from those involved in the investigation. Besides the obvious solutions to the accident problem, alternate solutions should also be discussed and considered. These may be immediate actions, midterm, long-term, or ongoing actions.

EVALUATE THE OPTIONS

Each option should now be considered and evaluated. The options should be evaluated without bias, and the best solution, the one that will fix the problem most effectively, should be chosen. Some remedies may be a combination of different proposed options. Some solutions to accident problems may be long term and some may be short term, so the best solution that will prevent further similar accidents from happening must be chosen. Solutions based on emotions will seldom solve the accident problem. Solutions that call for massive capital expenditure will seldom be approved by management. The test for the solutions posed is, *will they solve the accident problem?*

SELECT AN OPTION OR OPTIONS

Once all the alternate solutions have been evaluated and discussed, the ideal solution, or solutions, must now be agreed on. This may involve discussing the various options and an evaluation and ranking process to select the ideal solutions. This solution must be clearly defined and where necessary, implementation strategies, costs, and timelines should be a part of the strategy.

DOCUMENT THE AGREED SOLUTIONS

It is important for the investigator(s) to document the accident remedial measures on the accident investigation form, and, if necessary, to compile a separate document in the form of a management action plan, for the implementation of the agreed remedies. The action plan must clearly state:

- What must be done to prevent a recurrence of a similar accident in the future?
- Who is responsible for which actions to be taken?
- By what date, or how often are the actions to be undertaken?

APPLY THE SOLUTIONS

The action plan proposed by the investigation should now be implemented according to the timeline. It is advisable to implement the action plan activities as soon after the accident as possible, to reduce the risk of a similar event occurring. All action plans should be in writing, and must clearly state positive steps to be taken, who is responsible for those activities, and timeframes for completion and accomplishment of the set objectives.

FOLLOW-UP

Using the action plan as a guide, a follow-up should be done to ensure that the actions have been taken and have been successfully completed. This may involve a re-inspection of the accident site to confirm that changes indicated in the action plan have been made. It may also call for a reviewing of administrative controls if they were part of the proposals. In that case, applicable records, standards, and procedures may need to be examined to confirm the completion of the recommendations. Discussions with employees, witnesses, and health and safety committees should indicate if the actions have been implemented and if they have proved effective.

THE FISHBONE DIAGRAM METHOD

This method uses a diagram that lists six main components that contribute to the accident. They are material, people, method, process, environment, and machinery. During the investigation, the contributing factors are listed and analyzed accordingly. The method is as follows:

- Identify the accident as the problem
- Brainstorm possible causes of the accident based on evidence gathered
- Organize the causes onto the bones of the fishbone diagram
- Determine which causes are most significant
- Perform a root cause analysis for the identified causes
- Propose solutions for the root causes

FAULT TREE ANALYSIS

Fault trees represent a deductive approach to determine the causes contributing to a designated failure, such as an accident. The approach begins with the definition of a top or undesired event, and branches backward through intermediate events until the accident is defined in terms of basic contributing events.

THE *WHY* METHOD

This is a simple method whereby the investigator asks the question "Why?" five times for each high-risk behavior and five times for each high-risk condition identified during the investigation. The answers will bring the investigation closer to the root causes. This is a form of root cause analysis and is the method used in the *Logical Sequence Accident Investigation Method* discussed in Chapter 14.

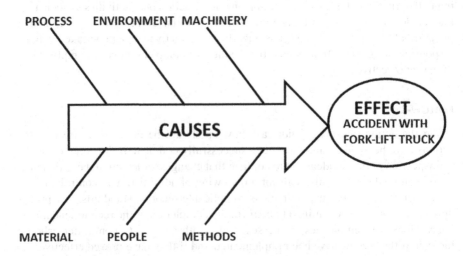

FIGURE 8.2   The Fishbone diagram.

WHO, WHAT, WHEN, HOW, WHERE, AND WHY?

This method enables the investigators to gather as much information as possible by asking the following questions:

- *Who* was injured, suffered ill health, or was otherwise involved in the adverse event?
- *What* happened at the time of the accident?
- *When* did the accident occur?
- *How* did the accident occur?
- *Where* did the accident occur?
- *Why* did the accident occur?

THE FIVE-STEP PROBLEM-SOLVING TECHNIQUE

This is a problem-solving method often used in quality control processes. The five steps are as follows:

- Define – it is important to define the specific accident problem. The more specific the accident problem is defined, the greater the chance of solving the problem.
- Measure – when the accident problem has been defined, decisions should then be made about additional measurements required to quantify the problem. Here the investigators should not only look at what happened but also what *could* have happened.
- Analyze – once the measuring stage has defined the accident proportions and probabilities, data is then collected from various sources and analyzed.
- Improve – after measurements have been taken and analyzed, possible solutions can now be developed to prevent the event from recurring in the future.
- Control – after the implementation of the solutions a follow-up must take place to ensure that the solutions have fixed the problem.

THE BRAINSTORMING METHOD

Group brainstorming, if done properly, can promote creative thinking, bring a team together, and help formulate solutions to specific accident problems. Brainstorming is a method accident investigation teams can use to generate ideas to solve accident problems. In controlled conditions and a free-thinking environment, teams approach the problem by asking "How could this accident have been prevented?" It encourages team members to come up with thoughts and ideas which are original and creative.

Brainstorming encourages out-of-the-box thinking and wild ideas presented should not be judged, nor criticized until the brainstorming is completed. All ideas are recorded on a chart and a voting process selects the best solutions to the problem.

PROBLEM-SOLVING SKILLS

Accident investigators need some problem-solving skills to help them define, analyze, and find an effective solution to the accident-causing problem.

COMMUNICATION

The investigator must be a good communicator, as he or she has to conduct accident witness interviews and hold discussions with other members who may be on the investigation team. Interaction and communication may also be required by bringing in experts and other outsiders who participate in the investigation.

ACTIVE LISTENER

Because of the amount of information related by witnesses, the investigator must be an active listener. This entails letting the witnesses relate the information without interruption. The investigator must listen carefully while noting all the details of the witnesses' recount of what they saw. Strategic questioning techniques should be used by the interviewer to encourage the witness to give as much information about the accident as possible.

ANALYTICAL ABILITY

Because of the information gathered after an accident, the investigator must be able to analyze the information, develop a clear picture of what events took place, how they took place, and what the causes were. This calls for basic analytical ability.

RESEARCH ABILITY

Investigators may need to research information concerning certain processes, chemicals, or other information relating to the accident. They should have some ability to do this research and apply this to the investigation.

DECISION MAKING

Once possible solutions to the accident problem have been proposed, the investigators and other members of the team will have to decide on the ideal solutions to implement and to prevent this accident recurring or a similar accident happening. This calls for basic decision-making skills.

TEAM BUILDING

If the investigation has been done by a team, working with the team is of vital importance. This calls for good team work, good communication, group participation, and excellent interpersonal relationships with the team.

CONCLUSION

Accidents should be treated as problems and be investigated using accepted problem-solving techniques. These methods are effective and help focus the investigation on the problem rather than on the person.

9 Rules of Accident Investigation

OBJECTIVE

Accident investigations must be approached with an open mind. Preconceived ideas, or prejudices, will skew the investigation and hamper a positive and accurate outcome. Investigators must not be swayed by circumstantial factors or any form of prejudgment. They should approach the investigation with an open and unbiased approach.

DO FACT-FINDING – NOT FAULT-FINDING!

The golden rule of accident investigation is *Accident investigation is fact-finding and not fault-finding.* In other words, use accident and near-miss incident investigation systems and methodology to get to the facts, the root causes of the accident, and not to find fault. Finding fault will not help in identifying and removing the causes of accidents, and it will not help solve the problem, nor will blaming the workers.

DELVE DEEPER

The investigators must delve deeper into the accident, and look beyond the apparent reasons to the underlying causes of the accident. All related past events should be recalled and examined. Near-miss incidents similar to the event being investigated should be studied and the circumstances collected as evidence.

LOOK AT BOTH SIDES OF THE COIN

In an accident investigation, both the employee's actions and management controls should be examined, analyzed, and recommendations proposed. The investigation cannot merely focus on one side of the coin, as this does not give a true reflection of the problem. Most accidents follow a series of breakdowns and the errors cannot simply be explained away, as they are often a series of blunders.

MORE INFORMATION

Often more information surrounding the event is required to ascertain all the facts. As much information as possible should be obtained. Gathering this information may require extensive research, fact-finding, as well as a close examination of past accident records, training records, maintenance records, witness interviews, and a thorough site inspection.

DOI: 10.1201/9781003220091-11

FACE VALUE

Investigators cannot accept all the answers given to them at accident review interviews. All information should be cross-checked. Facts and statements must be confirmed and correlated with other evidence and other witnesses. The passage of time tends to distort witnesses' recall of events, and this must be taken into consideration as well.

WITNESSES

Obtaining information and facts from witnesses to the accident is of vital importance. The investigator needs to obtain statements from witnesses to the event. Eye witnesses, employees who saw the event, are of prime importance. Other witnesses could include employees who do the same task, or who have operated or used the same equipment previously. Important witnesses could be those who supervised the employee, those who were involved in the training of the employee, or who in some way are connected to the employee.

PREJUDGE

Investigators should not prejudge the outcome of the investigation, and should not jump to conclusions. They should be briefed to gather the facts, and focus on the causes of the problem.

JUMPING TO CONCLUSIONS

The appointed investigator(s) should be briefed and cautioned not to jump to conclusions during the investigation, nor to base their findings on emotions, but rather focus on the facts of the event and the direct and indirect causes.

POTENTIAL FOR LOSS

It is important to determine and assess the potential of the mishap as soon as possible. The losses may appear apparent upon inspection of the site, but the potential of what could have occurred under slightly different circumstances should be determined. The accident may have had the potential to have caused more serious injury, multiple injuries, or more severe property damage. A mini risk assessment (Figure 9.1) of the event can be done by applying the following criteria: "It's not *what* happened, but what *could* have happened!"

LOOK BEYOND THE INJURED PERSON

After an accident, there is a tendency for investigators to focus on the injured person or persons. In some instances, the injured worker is interviewed while still receiving medical treatment or being transported to the hospital. To conduct an effective investigation, the investigators must look beyond the injured person. They should respect

LOSS POTENTIAL	PROBABILITY OF RECURRENCE
❑ LOW	❑ LOW
❑ MEDIUM	❑ MEDIUM
❑ MEDIUM-HIGH	❑ MEDIUM-HIGH
❑ HIGH	❑ HIGH
❑ EXTREME	❑ EXTREME

FIGURE 9.1   A mini risk assessment matrix.

the trauma of the injured party and rather focus on gathering information from other witnesses until the injured person has been stabilized, and is in a fit state to recall what happened. When interviewing an injured party, only the facts of what happened should be ascertained and opinions should be set aside.

PERCEPTION OF OTHERS

Considering the fact that others often see things differently, the investigator should take this into account and carefully weigh up the information from all witnesses and others involved. The different perceptions may cause the way in which something is regarded, understood, or interpreted, to differ from person to person.

COMMUNICATION SKILLS

In gathering information, the investigators must take other factors into consideration, such as language barriers, different cultures, and different levels of education, understanding, and experience. Some interviewees may give too much information, and some too little. Fear of "becoming involved" could also create a barrier to communication.

FACTUAL INFORMATION

What is most important, is to get as much information surrounding the event as possible. Getting all the information will allow the investigators to get the full picture. All sources of information should be pursued until the investigators are satisfied all pertinent information has been accumulated.

RUMORS

After an accident, rumors spread through the workplace quickly. Each employee adds their input to the story and eventually it becomes grossly distorted. Rumors are a source of unofficial communication in many organizations. To set the record straight, an initial accident report should be compiled and sent out to all employees and all sectors of the company. This initial report will state the true facts of what occurred, and although the investigation is not completed, it will state what happened, who was involved, and what the losses were. This will help stifle rumors and stories being spread about the accident.

TIME

The recollection of the accident, as well as the sense of urgency to find the accident causes, fades with time. Investigations should be started and completed in the shortest reasonable time. Dragging out the investigation leads to witnesses' recall of the event becoming distorted and less accurate. Valuable information and evidence could be modified, changed, or lost completely if the investigation is delayed. Follow-up actions to prevent a recurrence of a similar event must be time related, and a follow-up should ensure that the recommendations are timeously completed.

CONCLUSION

An accident investigation can open a can of worms. It should be fair, look at both sides of the coin, and not be a fault-finding exercise. An accident should be viewed as an opportunity to investigate, identify where the safety management system failed, and set up the necessary control measures, to prevent it from happening again. By following these rules, an accident investigation will become a meaningful exercise that will be beneficial in preventing similar future events.

10 Designing an Accident Investigation Report Form

A GUIDE

The accident investigation report form, often referred to as an *investigation form*, is a crucial part of every investigation and is the backbone document. This form could be in hard copy, or electronic format, either on a laptop computer or a tablet. The form must be compiled in such a way that it guides the investigator through the process. It serves as a checklist, helps the investigator identify both the immediate and root causes, has a space for remedial actions, and provides a one-form summary of the event.

CHECKBOXES

Where possible, the form must provide options to be selected with checkboxes. This will eliminate a lot of guesswork on the part of the investigator, and also provide a comprehensive investigation. Checkboxes will eliminate lengthy written responses, and act as prompts for the investigator.

TYPE OF EVENT

The form must provide a description of the type of downgrading event being investigated. This could range from a serious injury accident to a property damage accident, fire or collision, or near-miss incident. The event can easily be described by selecting the correct box on the form.

INJURY SEVERITY

The form must also provide a checkbox for the severity of the injury to be indicated. This may just be an estimate at first and may later be modified after hospital diagnosis. The two categories should be the nature of the injury and the body part injured. This information is also necessary for later injury analysis processes.

GENERAL INFORMATION

The following should be included in this portion of the investigation form:

- Injured or Involved Persons
  Provision must be made on the form for the name, or names, of the injured person or persons involved in the event. Where possible, employee numbers must be included.

DOI: 10.1201/9781003220091-12

- Date and Time

   The date and time of the accident, as well as the date reported, must be entered in the form in the spaces provided. This information is of vital importance and must feature on the form. The date of the investigation should be the same as the date of the event. The time the shift started is also relevant to the investigation and should be recorded.
- Task at Time of Accident and Task Experience

   The task being carried out when the accident occurred must be described as well as the experience of the persons carrying out the task. This is important information to ascertain exactly what experience the involved employees had in carrying out the task. If the accident occurred during overtime work, this should also be stated in the relevant block provided (Figure 10.1).

General Information of Employee(s) Involved			
Employee Name:	Employee #		Exact Location:
Occupation:	Time Event Occurred:	□am□pm	Task At Time:
Date Event Occurred:	Time Reported:	□am□pm	Years / Months Experience At Task:
Date Reported:	Time Shift Started:	□am□pm	Date Investigated:
Department:	Shift: Crew/Team:		
Contractor:	Overtime: □No □Yes If yes how many hours?		
Section:			
Supervisor Name:	Employee #: Tel:		Date Signed By Supervisor:
Manager Name:	Employee # Tel:		Date Signed By Manager:

FIGURE 10.1   General employee(s) information.

DAMAGE

Provision must be made on the form to indicate if there was damage and how severe the damage was. Initially, the cost may have to be estimated and further investigation will help determine actual costs. The organization must determine which damage events need to be investigated. Where a damage accident had the potential to cause injury under different circumstances, it should be investigated as a matter of course (Figure 10.2).

DAMAGE SEVERITY
□ No Equipment Damage
□ Damage < $5000
□ Damage > $5000
□ Extensive Damage
□ Business Interruption
□ Other

FIGURE 10.2   Ranking of the damage severity.

DESCRIPTION

Space should be provided on the form for a brief description of the event. In this space, the investigator would write a brief summary describing the event. This should be brief and concise and other, more detailed descriptions can always be included as annexures to the form.

ENERGY EXCHANGES

In older health and safety literature, this segment of the accident was incorrectly termed the "accident type." In fact, this is the type and source of exposure, impact, or energy exchange that takes place to cause the loss. The 16 main categories are listed below. Specific energy exchanges, under these main headings, will be identified during the investigation (Figure 10.3).

PHOTOS

Photos, sketches, and diagrams tell thousand words and must be included in the investigation report form. Some digital accident investigation forms allow photos to be pasted into the space provided on the form or hard copies can be attached in the space provided.

COSTING

Although not always possible to do the costing of the accidental losses before the investigation is completed, an estimate could be used that could be adjusted later at the conclusion of the investigation.

RISK ASSESSMENT

Bearing in mind that the outcomes of an accident are often fortuitous, and in many cases could have been worse. A risk assessment of what could have happened under different circumstances should be done. This should be provided in a block on the form. The two criteria should be the *loss potential* and *probability of recurrence* (Figure 10.4).

Exposure, Impact or Energy Exchange Type			
☐ Struck against	☐ Caught between	☐ Overexertion	☐ Caustics / Acid
☐ Struck by	☐ Slip	☐ Electricity	☐ Noise
☐ Caught in	☐ Fall from same level	☐ Heat	☐ Toxic substance
☐ Caught on	☐ Fall to lower level	☐ Cold	☐ Foreign object

FIGURE 10.3 Description of the exposure, impact, or energy exchange type.

Loss Potential	Probability of Recurrence
❑ Low	❑ Low
❑ Medium	❑ Medium
❑ Medium-high	❑ Medium-high
❑ High	❑ High
❑ Extreme	❑ Extreme

FIGURE 10.4   A mini risk assessment should be included in the form.

IMMEDIATE ACCIDENT CAUSES

An accident investigation also indicates who was involved at the time of the accident, and most importantly, is a means to identify the immediate causes in the form of high-risk behavior (unsafe or sub-standard act) and high-risk workplace condition (unsafe or sub-standard condition).

High-risk behavior (unsafe act) is a departure from a normal accepted or correct work procedure, which reduces the degree of safety of that procedure. A high-risk workplace condition (unsafe condition) is any physical condition that constitutes a hazard and has the potential to lead to an accident if not rectified.

Immediate (direct) causes of accidents are the causes in the chain of events, which are closest to the accidental contact. These immediate (direct) causes are the high-risk behavior and high-risk workplace conditions and, once determined, lead to the identification of the root or underlying causes in the form of *personal factors* and *job factors*.

QUICK REFERENCE

To act as a quick reference for the investigator, these high-risk situations should be listed on the investigation form so that they guide the investigator and provide a selection of checkboxes from which to identify both the acts and conditions, prevailing at the time of the accident (Figures 10.5 and 10.6).

HIGH-RISK BEHAVIOR

Although there are numerous high-risk behaviors or acts, the following are considered to be some of the most common high-risk acts:

- Operating without authority
- Working/operating at an unsafe speed
- Making safety devices inoperative
- Removing safety devices
- Using defective equipment
- Failure to warn
- Failure to secure

HIGH-RISK BEHAVIORS

☐ Operating equipment without authority
☐ Working / operating at improper speed
☐ Making safety devices inoperative
☐ Removing safety devices
☐ Using defective equipment
☐ Failure to warn
☐ Failure to secure
☐ Failure to use PPE
☐ Using equipment improperly
☐ Improper loading
☐ Improper placement
☐ Improper lifting
☐ Improper positioning
☐ Working on or servicing equipment in operation
☐ Horseplay
☐ Not following procedure
☐ Other (specify)

FIGURE 10.5 Checkboxes for high-risk behaviors.

HIGH-RISK WORKPLACE CONDITIONS

☐ Inadequate guard or barrier
☐ Inadequate or improper PPE
☐ Defective tools / equipment / materials
☐ Noise exposure
☐ Inadequate ventilation
☐ Hazardous environment
☐ Congestion/restricted / overcrowded area
☐ Fire and explosion hazard
☐ Poor housekeeping / unsafe arrangement
☐ High or low temperature exposure
☐ Radiation exposure
☐ Inadequate warning system
☐ Inadequate lighting / excess illumination
☐ Other (specify)

FIGURE 10.6 Checkboxes for high-risk conditions.

- Failure to use personal protective equipment (PPE)
- Using equipment improperly
- Improper loading
- Improper placement
- Improper lifting
- Improper positioning
- Working or servicing equipment in operation
- Distracting, teasing, or horseplay
- Not following procedures

HIGH-RISK WORKPLACE CONDITION

There can be numerous high-risk workplace conditions but the following are considered to be the basic categories of high-risk conditions:

- Unguarded, absence of required guards
- Inadequate guard or barrier
- Defective, rough, sharp, or cracked tools/equipment/materials
- Unsafe designed machines or tools
- Unsafely arranged and poor housekeeping
- Inadequate lighting, sources of glare
- High- or low-temperature exposure
- Inadequate ventilation
- Noise exposure
- Radiation exposure
- Unsafely clothed, no PPE
- Unsafe process
- Inadequate warning system
- Fire and explosion hazard

These immediate accident causes will be explained in more detail in later chapters.

ROOT CAUSES

Once all the immediate causes have been identified, a root cause analysis takes place to uncover the root or underlying causes. Once again, the form should list the main categories of root causes to guide the investigator. While the guide provides prompts as to the root causes, the investigator should show evidence from the investigation to quantify the root causes (Figures 10.7 and 10.8).

PERSONAL PROTECTIVE EQUIPMENT

To complete a descriptive investigation report, provision should be made to indicate what PPE was being worn at the time of the accident. Again, this can be a small checkbox included in the form.

PERSONAL (HUMAN) FACTORS

☐ Stress

☐ Lack of knowledge

☐ Lack of skill

☐ Inadequate capability

☐ Improper motivation

☐ Physical problem

☐ Other (specify)

FIGURE 10.7   The first category of root causes – Personal factors.

JOB (WORK / ORGANIZATIONAL) FACTORS

☐ Inadequate supervision or leadership

☐ Inadequate engineering

☐ Inadequate purchasing

☐ Inadequate maintenance

☐ Inadequate equipment or tools

☐ Inadequate ergonomic design

☐ Wear and tear

☐ Abuse or misuse

☐ Inadequate work standards

☐ Other (specify)

FIGURE 10.8   The second category of root causes – Job factors.

WITNESSES

A box should allow for entering the names and employee numbers of witnesses to the event. This will provide an immediate record of the list of witnesses for future reference (Figure 10.9).

WITNESSES	
Name:	Employee #:
Name:	Employee #:
Name:	Employee #:
Name:	Employee #:
Name:	Employee #:

FIGURE 10.9   The list of witnesses.

REMEDIES (CONTROL MEASURES)

The most important inclusion in the investigation form is the space provided to list actions to be taken to prevent a recurrence of the same or a similar event. This space should provide at least five lines for remedial measures recommended by the investigator. Included in this list should be a management action plan stating *who* must do *what* and by *what date* it should be completed. A checklist for the date on which the control measures have been completed should be included so the investigators have a follow-up mechanism to close the loop (Figure 10.10).

FOLLOW-UP

Once the remedies have been recommended and implemented, it is of vital importance to carry out a follow-up exercise to ensure they have been implemented and are effective. If they do not solve the problem, then other remedies should be sought. A sign-off check sheet can be included in a space provided on the form to insert the date of completion of the remedial actions.

SIGNATURES

Once all the remedial measures have been implemented and monitored for effectiveness, the form should be signed off by the investigator. In the case of a group investigation, the lead investigator signs the form. The manager of the section where the

CONTROL STEPS TO PREVENT A RECURRENCE	WHO MUST DO IT?	DATE TO BE COMPLETED	DATE COMPLETED
1.			
2.			
3.			
4.			
5.			
FOLLOW UP TO MONITOR THE EFFECTIVENESS OF THE CORRECTION ACTIONS			DATE COMPLETED
1.			
2.			
3.			
4.			
5.			

FIGURE 10.10   Control steps to prevent a recurrence and management action plan.

accident occurred then signs off on the investigation. The report form is then routed to the one-up manager for signature and then the top manager of the organization. Executive management should sign off on all serious accident investigations. It is important that each manager adds comments next to their signatures.

INVESTIGATION CLOSE-OFF

Final close-off of the investigation is done by the safety department. They should be the last to sign off indicating they are satisfied with the investigation and outcomes.

CONCLUSION

The key to a successful investigation is an accident investigation form that leads the investigators through the process. It should be a self-prompting document that gathers vital information by using checkboxes. This type of form reduces the amount of writing needed to record information and contains self-prompts which just need checking on the form to record details. It should include a risk assessment, costing section, and an action plan to implement remedial measures.

11 Accident Site Inspection

THE STARTING POINT

The *Logical Sequence Accident Investigation Method* calls for the investigation to start at the accident site, where the loss occurred. The aim of the site inspection is to get the facts as to which employees and what equipment were involved. It further provides information appertaining to the positioning of the equipment, tools, and machinery, and what the environmental conditions were. These would include illumination, visibility, noise, and the general work situation. The site inspection allows the investigator to record the losses produced by the adverse event, as well as the impacts (exchanges of energy) that caused those losses.

SECURING THE SITE

Immediately upon the occurrence of an accident, the area manager or supervisor should take charge of the scene, ensure the site is safe, and ensure that the necessary medical attention and evacuation are initiated. Steps should be taken to ensure that no secondary accidents occur. The site must be secured and made safe.

INVESTIGATORS' CAUTION

When visiting an accident site, the investigators should be made aware of the safety requirements of the area. If they have not been to the site before, the organization may require them to undergo standard safety induction training before moving onto the site. Where applicable, outside investigators should be issued with personal protective equipment and should be accompanied at all times.

SOURCE OF INFORMATION

The prime source of information about the accident is the physical site where the accident occurred. There may be more than one physical location. The site will form the canvas on which the investigators will paint a picture of the event.

Because of the importance of the evidence gathered at the site, it is important to protect the area from external interference so that it cannot be disturbed. Once immediate action has been taken to render assistance to injured persons, the site should be made safe from secondary accidents and thereafter should be roped off and secured. Only once all the evidence has been gathered should the site be released for normal production to continue.

DOI: 10.1201/9781003220091-13

PHYSICAL EVIDENCE

The site will offer up physical evidence that will help with the gathering of information surrounding the accident. This evidence may be in the form of parts of machinery or equipment, tools and devices used, such as ladders, electrical equipment, or other items.

POSITION

The position of items at the accident site is a vital source of information. It is, therefore, imperative that photographs be taken of the accident site as soon after the event as possible, for later scrutiny. Unfortunately, in many cases, the site is altered before the investigation commences which leads to a loss of vital site-related evidence. While making the site safe to prevent additional loss, damage, or injury, vital physical evidence should be noted or photographed before the site is altered to ensure no evidence is destroyed.

GATHERING EVIDENCE

Evidence is vital for an effective accident investigation. According to the Canadian Centre for Occupational Health and Safety (2019b),

> Before attempting to gather information, examine the site for a quick overview, take steps to preserve evidence, and identify all witnesses. In some jurisdictions, an incident site must not be disturbed without approval from appropriate government officials such as the coroner, inspector, or police. Physical evidence is probably the most non-controversial information available. It is also subject to rapid change or obliteration; therefore, it should be the first to be recorded. Based on your knowledge of the work process, you may want to check items such as:
>
> - positions of injured workers
> - equipment being used
> - products being used
> - safety devices in use
> - position of appropriate guards
> - position of controls of machinery
> - damage to equipment
> - housekeeping of area
> - weather conditions
> - lighting levels
> - noise levels
> - time of day (Website 2021)

ENVIRONMENTAL CONDITIONS

The environmental factors at the site are important indicators and should be recorded during the site visit. Some of the questions investigators should ask are as follows:

- What were the weather conditions?
- Was it too hot or too cold?
- What was the humidity?

- Was noise a problem?
- Was there poor housekeeping?
- Was the illumination adequate?
- Were toxic or hazardous gases, fumes or dusts, present?

PHOTOS AND SKETCHES

Getting the facts may involve

- taking photographs
- making sketches
- taking measurements
- examining items
- reviewing standards
- reviewing procedures
- examining the past experience as well as the accident history of employees.

It is important to keep an open mind during the gathering of these facts, as conclusions should not be jumped to, but should be derived from the facts.

According to the Health and Safety Executive (HSE), publication *Investigating Accidents and Incidents* (2004c),

> Discovering what happened can involve quite a bit of detective work. Be precise and establish the facts as best you can. There may be a lack of information and many uncertainties, but you must keep an open mind and consider everything that might have contributed to the adverse event. Hard work now will pay off later in the investigation.

(p.14)

DIAGRAMS

If pertinent, any diagrams or production schedules should also be gathered while at the site. Flow diagrams and workplace layout plans may need to be requested for further pertinent, relative information.

PEOPLE

People are a vital source of information. The different categories of people evidence are as follows:

INJURED PARTY

Traditional investigation techniques viewed the injured person or persons as the most important witnesses. While they may be important witnesses, they should not be regarded as the most important witness, as there are many others who may contribute much to the investigation. When interviewing the injured party, one should be mindful of their trauma and not attempt to interview them until they have been treated and are in a stable state, and able to answer questions.

EXPERTS

In many cases, certain experts in a specific field or industry may need to be involved in the investigation. Experts qualified in these specialized fields are the best persons to investigate these accidents. Other specialists may be called on to assist with the investigation, depending on the nature of the industry and process. The advantage of subject experts being recruited to assist in the investigation is their extensive experience and knowledge of the specialized area. They are also, outside, impartial investigators, not influenced by internal company politics. They can provide much-needed specialized information regarding the event.

WITNESSES

Witnesses to the event are vital in helping to ascertain what happened and how it happened. Some may have seen the event unfolding, others may have heard something happening, and some could have been involved in the event. Bystanders should also be interviewed as a source of evidence, as their observations are important.

Names and phone numbers of witnesses should be taken at the scene where possible. A positive approach should be taken with witnesses and they should be convinced that the investigator only wants to know "what happened." Interviews should take place at a suitable venue away from the site, and should not be confrontational.

DOCUMENTED EVIDENCE

Documented evidence appertaining to the accident is vital as it will help fill in the gaps and answer questions raised by the investigators. These documents could include:

- Employee training records
- Medical records
- Maintenance records
- Previous accident investigation reports
- Near-miss incident reports
- Inspection reports
- Risk assessments
- Written job safe work procedures
- Photographs/diagrams/maps
- Work permits issued

SITE REVISITED

In some cases, it may be necessary to revisit the site on more than one occasion. Sometimes, evidence is overlooked during a single site visit. Revisiting the site may well disclose some evidence or situation that at first was not noticed.

WORKPLACE STANDARDS

While the investigators are primarily focused on the specific site of the accident, it may be opportune for them to examine the workplace surrounding the accident site. Does the housekeeping in the rest of the area appear to be in order? Are walkways and aisles demarcated and free from obstructions? Are safety signs in place indicating the position of firefighting equipment, emergency exits, and other information? These observations will indicate the state of the health and safety system in operation and give the investigator an idea of the prevailing safety culture.

CONCLUSION

The accident site and site inspection are vital to an effective accident investigation. Investigators should always visit the site and begin their investigation with the evidence presented by the site. Many clues to the accident causation can be gathered at the accident site, and therefore the accident site inspection is a vital kick-off point of any investigation. Only once all the impacts or exposures have been identified by the losses noted at the site, can the investigators move their attention to the immediate causes.

12 Interview Guidelines and Interviewing Techniques

INTRODUCTION

Witnesses are vital to an accident investigation, as they are there before, during, and after the accident. They possess first-hand knowledge of what happened, and they are usually experts in the subject relating to the accident. It is important to interview witnesses as soon as possible after the event to ensure their recollection of the accident is still accurate. If a person (or persons) was injured, it is also vital to get their testimony. Before interviewing them, ensure that they are not under duress due to pain and suffering, and are in a position to give information. If not, wait until the attending medical practitioner gives clearance for the interview. Always respect the injured person's injuries and accompanying trauma.

INTERVIEW VERSUS INTERROGATION

- An *interview* is a formal meeting in which one or more persons question, consult, or evaluate another person. It is a meeting or conversation in which an accident investigator asks questions of one or more persons, from whom information and facts are obtained concerning an accident.
- An *interrogation* is an intense formal and official questioning session, where witnesses are questioned formally and systematically. It consists of tough direct questions. This technique is usually used by police officers to obtain information.

The most important thing to remember during post-accident interviews of persons who are involved and are witnesses to the accident is that it is *not an interrogation*. The interview is not a forum to subject the interviewees to police or military-type questioning techniques. The interview is a fact-gathering forum and should be conducted as such.

SUITABLE LOCATION

Post-accident interviews should be conducted in a quiet, private, comfortable location away from sources of disturbance that is free from disruption. Seating should be comfortable and refreshments should be available. Frequent breaks should be taken, and more than one interview session may be needed to conclude an interview.

DOI: 10.1201/9781003220091-14

INTERVIEW GUIDELINES

If an employee is being interviewed and is concerned that the interview may result in disciplinary action being taken against him or her, they may request shop floor (Union) representation. If a representative is requested, the interview should be terminated until representation is obtained.

The interviewer should prepare open-ended and possible follow-up questions for witness interviews. In some instances, the witness may have to be taken to the accident site to indicate certain features and details to the interviewer.

Interviewers should show empathy during interviews and make no attempt to fix blame or find fault. Be objective during the discussions and don't form advanced opinions. Avoid using questions that lead the witness. When the witness finishes the initial explanation, ask questions to fill in any gaps.

Summarize your understanding with the witness after the interview so that there is a complete understanding of the statement. At the conclusion of the interview, express sincere appreciation to the witness for helping in the investigation, and be sure to record the interview information accurately.

All interviews should be recorded. The name, work address, phone number, date, and signature of the interviewer should be included in the document. The statement should be read back to the interviewee so that he or she understands and agrees fully with its contents. Where possible, a witness signature on the statement should be obtained.

TYPES OF WITNESSES

There are different types of witnesses to any event and the main classifications are as follows:

- A *primary* witness is a person with first-hand knowledge of an event, and who testifies to that knowledge during an accident investigation.
- An *eyewitness,* also known as a percipient witness, is one with knowledge obtained through his or her own senses, such as visual perception, hearing, smell, or touch.
- A *hearsay* witness is one who testifies about what someone else said or wrote.
- An *expert witness* is one who allegedly has specialized knowledge relevant to the accident, which knowledge purportedly helps to either make sense of other evidence, including other testimony, documentary evidence, or physical evidence.
- A *reputation witness* is one who testifies about the reputation of a person or organization, when reputation is material to the accident investigation. Such a witness is someone who testifies to a person's interactions and personality, so their role in the event is clearly defined. This could apply to work teams and groups as well.

VICTIMS

In some instances, the injured person(s) could be a victim of a system failure that led to the accident. In any instance, the injured person or persons suffer physically, emotionally, and financially, as a result of an injury or occupational disease. This should be remembered during the interview process. Even if the injured person(s) did commit unsafe practices that contributed or may have contributed to the accident, they should not be treated as "guilty" parties during the interviews. The interview is to gather all the facts, not to place blame on individuals.

COLLEAGUES

It is a difficult situation when workers have to give information about their colleagues' actions at a post-accident witness interview. Many would feel they are letting their colleagues down or telling stories, or "ratting" on them, even if the friends did nothing irregular. A positive climate should prevail in the interview location, interviewees should be reassured that no victimization will take place, and that the interview's purpose is to gather the facts so that action can be taken and similar accidents can be prevented.

EXPERTS

If experts are called in to help the investigation, they should also follow the same interview protocols and treat interviewees with maximum respect. Under no circumstances should the interview evolve into an interrogation session.

QUESTION TECHNIQUES

To gain the most meaningful information concerning the accident from a witness, correct questioning techniques must be used. Asking the wrong type of questions may lead to the witness waffling and missing the point of the question and wasting time. The most effective technique is to use a variety of question techniques to derive as much information as possible.

Understanding the specific types of questions helps achieve better answers and gather vital information, but will also help avoid misleading the witness, or worse, preventing a breakdown in communication.

TYPES OF QUESTIONS

The nine basic question types which will be discussed are as follows:

- Open
- Closed
- Loaded
- Probing

- Leading
- Funnel
- Recall
- Process
- Rhetorical
- Mirrors

Open Questions

To find out more information concerning the event, open questions are frequently used. These questions require a little more thought and generally encourage wider discussion and elaboration. They encourage the interviewee to explain and elaborate. For example, "What were you doing at the time of the accident?"

Closed or Polar Questions

Closed, or polar questions, usually invite a one-word answer, such as yes or no. They are useful in gathering a quick, direct answer. For example, "Did you see the accident happen?" or, "Were you on the deck when it happened?" These questions could also include answers to factual, or multiple choice questions, such as "In what department do you work?" or "How long have you been a crane operator?"

These questions are good as icebreaker questions to help the interviewees relax and become comfortable with the interview process.

Loaded Questions

Loaded questions contain an assumption about the person being interviewed. They're often used by journalists and lawyers to trick the interviewee into admitting a fundamental truth they would otherwise be unwilling to disclose.

For example, the question, "Have you always ignored the safety rules in the welding shop?" This question infers the respondent does ignore the safety rules. Irrespective of the answer, the interviewee may admit to having ignored some safety rules in the past and therefore may admit to having breached safety rules, as it is not easy to spot the trick in the question.

Probing Question

Probing questions are useful for gaining clarification and encouraging others to tell more information about the accident being investigated. They help to get information from witnesses who may be reluctant to divulge information, or who are shy. Probing questions are a series of questions that dig deeper and provide a fuller picture. For example, "Were you wearing gloves at the time of the accident?" or "Who is responsible for checking the machine guards?"

Leading Questions

These questions are designed to lead the respondent toward a certain desired positive or negative route. These questions should be asked carefully to avoid confrontation. Leading questions lead the witness to a positive or a negative response, which helps clarify the situation that led to the event. For example, "Do you think management do enough for safety here?" or "Is your workplace as safe as you would like it to be?"

Leading questions can prompt either a negative or positive response. A more balanced answer can be obtained by a question such as "How do you participate in the safety program here?"

Leading questions could also involve an appeal that's designed to coerce the interviewee into agreeing with the interviewer. For example, "Safety is a priority here isn't it?" This encourages the witness to say yes, or conversely a question such as "Do you think safety standards can be improved here?" may be answered with a negative response.

Funnel Questions

These questions begin broadly before narrowing to a specific point, almost like a funnel, and can also be used to diffuse the tension of a witness interview.

When interviewing a witness, we usually begin with specific, closed questions, such as "What's your name?" or "Where do you work?" Once the ice is broken, the interviewer then broadens out into more open-ended questions, such as "Which part of the accident did you witness?"

These questions can be used in reverse. The interviewer begins with a broad question before focusing on something specific. This technique is used to gather as much information as possible.

By asking the witness being interviewed to go into detail about their issues, distracts them from their anger and gives the information needed to offer a solution. This helps to calm the witness down and gives him or her the impression that something positive is being done.

Recall Questions

Recall questions require the recipient to remember and recall facts about the accident. For example, "When did the accident happen?" or "How many workers were injured?"

Process Questions

Process questions can be used to test the witnesses' depth of knowledge about a particular topic. For example, "What do you think went wrong and caused the accident?" This type of question invites the witness to give their opinion on the event.

Rhetorical Questions

These questions are statements phrased as questions to make the conversation more engaging for the witness, who is drawn into agreeing with the interviewer. These questions don't really require an answer. For example, "Isn't this a great company to work for?" or "I was very impressed with the standard of housekeeping in this plant." Rhetorical questions are asked to create a dramatic effect, or to make a point, rather than to get an answer.

Mirror Questions

Mirror questions involve restatements and reflection of what the other person has just said. These types of questions are used to encourage others to speak. Mirrors are nondirective. They are intended to encourage the person being interviewed to continue to add detail to what they have said, without being influenced to go in a specific direction. They show interest in what the witness has to say and can help minimize misunderstandings and increase rapport. For example, the interviewee could say, "I was nearly involved in a similar accident many years ago." The interviewer would mirror this and say, "I'm sure you are glad that you were not injured." The interviewer mirrors or rephrases the statement to gain more information.

ACCIDENT RECONSTRUCTION

In some cases, it may be necessary to reconstruct the accident with the help of witnesses. A word of warning do not re-enact the accident so convincingly that the same event recurs, and further injury and damage are incurred. Explain this to the persons involved in the reconstruction. Ask witnesses to demonstrate what was happening at the time of the accident. Do everything in slow motion and remind them not to injure themselves, or others while re-enacting the accident. A video recording of this accident reconstruction should be made.

CONCLUSION

Witnesses are vital to an accident investigation, as they are there before, during, and after the accident. They possess first-hand knowledge of what happened, and they are usually experts in the subject relating to the accident. They are the most important source of information about the accident.

Post-accident interviews of witnesses are interviews, not interrogations. An *interview* is a formal meeting in which one or more persons question, consult, or evaluate another person. An *interrogation* is an intense formal and official questioning session, where witnesses are questioned formally and systematically. To gain the most meaningful information concerning the accident from a witness, correct questioning techniques must be used. Asking the wrong type of questions may lead to the witness waffling and missing the point of the question and wasting time.

13 Documentation Review

INTRODUCTION

In an investigation, a valuable source of information concerning the accident can be gained from documented actions, policies, standards, and records. An accident is indicative of one or more management controls failing. Documents indicating these controls should be reviewed as part of the investigation to determine where one or more were inadequate, nonexistent, or failed.

STANDARDS

The organization should have an established Occupational Health and Safety Management System (SMS) in place. The basic elements of the SMS are normally local legal health and safety requirements, upon which the organization has built an effective SMS. An effective SMS consists of a number of written safety standards, policies, and procedures. The standards applicable to the accident should be reviewed and compared for deviations that may have led to the accident.

PROCEDURES

After establishing what procedures were being undertaken at the time of the accident, the written procedure should be called for and scrutinized to establish where deviations may have occurred. Perhaps the procedure is inadequate, outdated, or has been changed without the employees being notified. The requirements of the procedure should be compared with what was found at the site. Not all tasks require a written procedure, but critical tasks with a high potential for loss should have written procedures in place.

Not a Quick Fix

In many cases after an accident, a procedure is written for the job or task being undertaken that led to the accident. This is not necessarily the solution. If the procedure does not address the root causes, then it will not fix the problem. Not all tasks require a written procedure or a method statement. Often these end up in a cupboard gathering dust and serve no purpose. If the task is a critical one, then a procedure should be available. This procedure should be derived from a task safety analysis (Job Safety Analysis), which is actually a task risk assessment. Frequent briefing sessions on the procedure should be held, and the procedure should be updated on a regular basis.

Case Study

Three employees were nearly seriously injured in a fall of ground event in an underground mine. After the accident, the supervisor produced an almost blank accident

DOI: 10.1201/9781003220091-15

investigation form after he had investigated the event. The only portion completed was the *remedial measures to be taken* section, where he had written, "The three employees must be disciplined for not following the procedure."

This was a serious near-miss incident. The employees could have been badly injured under different circumstances; therefore, emotions were high during the initial investigation. When a team of investigators reviewed the accident investigation report, they immediately called for the procedure. It took more than two days for the procedure to be found! Once it was made available and was being reviewed by the investigators, an operator from the same division noticed that the procedure was outdated, as it had been updated more than three years previously! This was an indication that the procedure was not readily available, and the outdated procedure had not been removed from the system. The employees were also not aware of the updated procedure.

According to the procedure, the investigators also found that the task was supposed to have been carried out under the direction of a supervisor experienced in the specific task. When the employees were questioned, the investigators were informed that the experienced supervisor had taken the day off, and that the task was under the direction of a supervisor who had never done the task before, and who knew nothing about the risks of the task. As a result, the employees were not disciplined.

If procedures are written for critical tasks, they should be readily available. They should be used regularly for training, and be updated on a scheduled basis.

The organization reviewed the procedure, improved it, and made the procedure available to the relevant employees. The improvement included the retraining of employees in the procedure and the revised scheduling of days off for key personnel.

Important Questions

- When was the procedure compiled?
- Does the procedure still apply?
- How often is this procedure updated?
- When was the employee trained in the procedure?
- When was the employee retrained?
- Is there written proof of this retraining?
- When last was a structured observation of the execution of the procedure carried out?
- Is this procedure discussed at toolbox meetings?

Safety Paradigm

In *Changing Safety's Paradigms*, McKinnon (2019) wrote

One of safety's paradigms is that after an accident a procedure must be written and the problem will be solved. That is a paradigm and needs to be shifted. Procedures must be developed in the correct manner, and only for critical tasks. Critical tasks are those tasks which have the highest risk. About 20 percent of tasks carried out at a workplace have 80 percent of the potential to cause accidents. Employees should be trained in these critical task procedures. The procedures need to be revised periodically, and retraining should be ongoing. Writing a procedure for every task is impracticable, and not always necessary. Writing procedures, and filing them in a cabinet as a cop-out should an accident occur, is defeating the objective. Writing a meaningless procedure after an accident will not fix the accident problem. Effective and thorough accident investigation, done correctly, will highlight the accident root causes, and remedial measures can be devised to eliminate them. Writing a procedure after every accident is missing the point.

(p.113)

TRAINING

Training records showing that employees have been trained, and retrained in the procedures should be examined by the investigators. Formal training is positive instruction, and simple verbal instructions are not effective safety training. Annual refresher training and other ongoing training should be provided and these records should be reviewed.

SAFETY INDUCTION TRAINING

A record of the employees' safety induction and annual refresher training should be available for scrutiny by the investigators. Safety induction training should be provided when an employee joins the company or when contractors are on site for more than a day. Long-term visitors to the site and consultants should also undergo safety induction training.

OTHER HEALTH AND SAFETY TRAINING

If other training in health and safety was provided, these records should also be scrutinized. A question to be asked by the investigator is "Is the training sufficient and is it effective for the task that the employee does?"

POLICIES

Organizations have basic policies in the workplace. A policy is a statement of intent and is normally implemented as a procedure or protocol by the organization. The investigator should establish what specific policies applied to the accident scenario to ascertain if the pertinent policies were applicable, or whether they were adhered to. In many instances, policies are not updated, not explained, or frequently ignored. Repeatedly circumventing policies without consequence is known as condoned behavior.

HEALTH AND SAFETY POLICY STATEMENT

Although it may not relate directly to the accident, this document will give the investigator a good idea of the safety intent of the organization. The policy should be current, prominently displayed, and be signed by the senior executive. Outdated policies that are not prominently displayed are a sign of a weakness in the SMS.

AUDIT REPORTS

If the organization's SMS has been audited recently, reviewing a copy of the audit report may shine light on weaknesses in the system that may have contributed to the accident. Investigators should request a copy of the last audit, and examine it for any evidence which could indicate a warning of a failure, which could have resulted in the accident.

INSPECTION REPORTS

Each work area should undergo a health and safety inspection on a regular basis. Sometimes, these inspections are done by the health and safety representative. Copies of the reports from these inspections are a good source of evidence for the investigator. They may indicate hazards that were reported by the inspector or other deficiencies. The last inspection report for the area where the event took place should be requested and examined. It will give a good idea of the physical conditions of the area at that time, and may help with the investigation.

MAINTENANCE RECORDS

Maintenance records and schedules could prove to be good sources of information for the investigation. This will depend on the machinery or equipment involved in the event. If a vehicle was involved, the maintenance records will give an indication of any defects or repairs undertaken, which may prove relevant to the event. If other equipment was involved, the checklist and registers will provide valuable information.

PAST EXPERIENCE

Was there a history surrounding the accident? In some instances, a similar accident may have occurred in the past and the root causes were never established and rectified. Is this a constant machinery failure, or does the event repeat itself every few years? These are some questions an investigator could ask. Records of previous events should be examined for possible trends.

PAST EVENTS

The investigator should also enquire if a similar event happened in similar types of industry and under similar circumstances. Similar events may have occurred in different countries, which could help identify what controls failed and what other common factors contributed to the accident.

For example, an underground locomotive fell down the mine hoist shaft and landed on a moving cage full of miners on their way to the surface. The collision caused the cables to snap which sent the cage and its occupants plunging down to the bottom of the shaft. More than 50 miners died in this tragic accident.

In a different country, many years before, an underground locomotive also fell down the main shaft of a mine, but did not hit the cage, as the cage was above the level from which the locomotive fell. In this instance, no one was injured, but the potential to cause multiple fatalities existed. Lessons learned from this high-potential adverse event can contribute to preventing similar accidents happening, before tragic consequences occur.

NEAR-MISS INCIDENTS

A record of previous near-miss incidents can give the investigator a good idea of the type and frequency of events that occurred in the past, with no loss. In some cases, near-miss incidents similar to the accident being investigated can expose warnings that were not heeded. If there was a near-miss incident, and no action was taken, then the probability of a recurrence of a similar event, with loss consequences, is high.

Similar past events should be examined to see if there is any trend, or if there were previous warnings in the form of near-miss incidents similar to the accident being investigated.

WITNESS STATEMENTS

The written statements from accident witnesses are vital documents, which will contribute much information about the event. These should be scrutinized along with other relevant documentation so that a clear picture of what happened is obtained.

INCIDENT RECALL

Recalling past events and situations such as high-risk acts and high-risk situations that occurred in the past is a good method of determining if the accident was just an isolated event. Were there warning signs of an accident about to happen? Recall sessions should be held on a regular basis, and employees should be encouraged to recall any deviations from safe practices, as well as accidents and near-miss incidents, whether they happened in the workplace, at another site, or during leisure time.

By reviewing incident recall records, the investigator will be able to obtain a clear idea of what occurred in the workplace previously. They may also indicate repeated high-risk behavior or condoned practices in the past. Since incident recall is such a vital part of an SMS, the incident recall process should allow for anonymous, no blame reporting.

CONCLUSION

Documents relating to the organizations' SMS are vital sources of information for the accident investigator. The SMS standards applicable to the accidents should be reviewed and compared for deviations that may have led to the accident. Training records and other pertinent records will deliver important facts relating to the event. Previous property damage accidents and near-miss incidents should be reviewed to ascertain if they gave warnings of high-risk situations that existed in the past. Safe work procedures should be tested to ensure they are current, being used, and are understood by the employees. A documentation review will help the investigators to complete the picture.

Section III

Investigating and Analyzing the Event

14 The Logical Sequence Accident Investigation Method (Cause, Effect, and Control)

INTRODUCTION

The *Logical Sequence Accident Investigation Method* is a simple, practical, and effective method of accident investigation.

- It deals with the causes, the effect, and the control phases of an accident.
- It follows a basic accident causation model and includes an immediate cause analysis as well as a root cause analysis.
- It starts with the tangible end results of the event and progresses to the intangible causes.

THE DOMINO EFFECT

A loss causation (accident sequence) model will be used to apply the *Logical Sequence Accident Investigation Method*. This accident sequence will be represented by a string of dominoes standing side by side. If the first one falls, a chain reaction occurs and all the dominoes fall. This is called the *domino effect* (Figure 14.1).

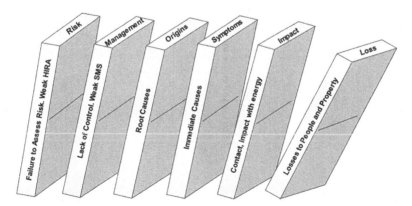

FIGURE 14.1 A basic accident (loss causation) domino sequence. (From McKinnon, Ron C. 2017. *Risk-based, Management-led, Audit-driven Safety Management Systems*. Boca Raton: Taylor and Francis. With permission).

DOI: 10.1201/9781003220091-17

A BASIC LOSS CAUSATION (ACCIDENT) SEQUENCE

In the basic accident sequence model, the event is triggered off by a failure to iden-tify certain hazards and carry out risk assessments, which in turn leads to a failure or weaknesses in the safety management system (SMS) controls. These weaknesses allow root causes to exist, which in turn lead to high-risk behaviors and high-risk workplace conditions. These behaviors and workplace conditions are the immediate causes that lead to the exposure, impact, or exchange of energy, which in turn cause losses to people, property, or both.

THE *LOGICAL SEQUENCE ACCIDENT INVESTIGATION METHOD*

Step 1

The first step in the *Logical Sequence Accident Investigation Method* is determin-ing the losses caused by the accidental exposure, impacts, or exchange of energy. These could be losses to people, property, equipment, product, the environment, or a combination of one or more (Figure 14.2).

CATEGORIES OF LOSS

The main categories of accidental loss are *direct* and *indirect* (insured and non-insured). Direct losses are normally visible at the accident site, while indirect losses may only become apparent during the later stages of the investigation. Losses could be losses involving people, property, equipment, materials, or the environment.

FIGURE 14.2   Examine all the losses caused by exposure, impacts, or exchange of energy.

Step 2

This step is to determine what exposures, impacts, contacts, or exchanges of energy took place to cause the losses. The exposure, impact, contact, and exchange of energy type could be:

• Caught between	• Fall to the lower level	• Heat
• Caught in	• Caught on	• Cold
• Overexertion	• Electricity	• Non-toxic substance
• Struck by	• Noise	• Toxic
• Slip	• Caustics/Acids	• Radiation
• Fall from the same level	• Foreign object	• Other (Specify)

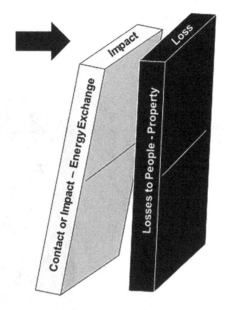

FIGURE 14.3   Determine the exposure, impacts, or contacts with energy sources.

Step 3

This step is to determine what high-risk behaviors and high-risk conditions led to the unexpected contact with a source of energy, which was above the threshold limit of the person or body. In some cases, there may be more than one impact and more than one high-risk act or high-risk workplace condition. This is the immediate cause analysis phase of the investigation.

The main categories of high-risk behaviors are as follows:

• Operating equipment without authority	• Using equipment improperly
• Working/operating at an improper speed	• Improper loading
• Making safety devices inoperable	• Improper placement
• Removing safety devices	• Improper lifting
• Using defective equipment	• Improper positioning
• Failure to warn	• Working on moving equipment
• Failure to secure	• Distraction, teasing, horseplay
• Failure to use PPE	• Not following procedures
• Other (Specify)	

The major categories of high-risk workplace conditions are as follows:

• Unguarded, absence of required guards	• Inadequate ventilation
• Inadequate guard or barrier	• Noise exposure
• Defective tools/equipment/materials	• Radiation exposure
• Unsafe designed machine or tools	• Unsafely clothed, no PPE
• Unsafely arranged and poor housekeeping	• Unsafe process
• Inadequate illumination, sources of glare	• Inadequate warning system
• High- or low-temperature exposure	• Fire or explosion hazard
• Other (Specify)	

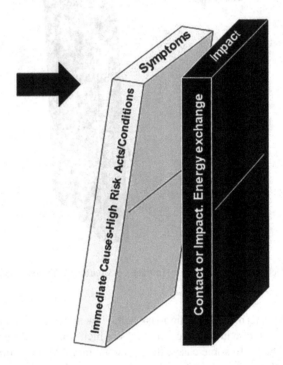

FIGURE 14.4 Identify the immediate causes (high-risk behaviors and conditions) that led to the exposure, impacts, and energy exchange.

Step 4

The fourth step is to single out each immediate cause and ask *why* it existed or took place. The question *why?* should be asked repeatedly until an underlying (root) cause, or causes, is established. The question must be asked repeatedly for each high-risk behavior, to determine the personal (human) factors, and for each high-risk condition, to determine the workplace environment (job) factors. This must be repeated until all the immediate causes have been analyzed for root causes. There may be more than one root cause for each immediate cause. The main categories of root causes are as follows:

Job (Workplace, Environment) Factors

• Inadequate engineering	• Inadequate maintenance
• Inadequate supervision or leadership	• Inadequate equipment or tools
• Inadequate ergonomic design	• Wear and tear
• Abuse or misuse	• Inadequate work standards
• Inadequate purchasing	• Other (Specify)

Personal (Human) Factors

• Stress	• Improper motivation
• Lack of knowledge	• Physical problem
• Lack of skill	• Other (Specify)
• Inadequate capability	

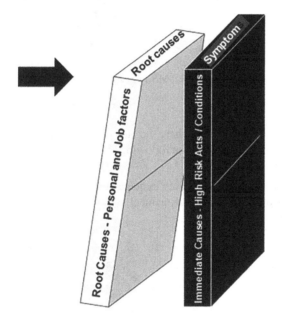

FIGURE 14.5 A root cause analysis is done by asking *why?* for each high-risk behavior or condition. The answers are the root causes.

Step 5

Once the root causes have been established, the SMS controls, processes, and procedures are examined to ascertain what caused the root causes to emerge. These may be:

• A lack of standards
• Non-compliance to standards
• A weakness in the SMS

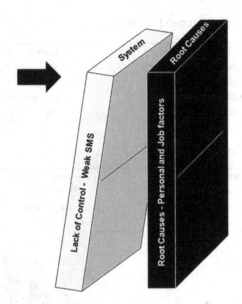

FIGURE 14.6 Identifying where the SMS failed or had weaknesses that gave rise to the root causes.

Step 6

The final step is to determine where the hazard identification and risk assessment (HIRA) system failed to identify the following risks:

- A non-existent health and safety management system
- A weak health and safety management system
- Inadequate control systems
- Non-compliance to standards

FIGURE 14.7 Ascertain where the HIRA system failed to identify the weakness in the SMS.

REMEDIAL RISK CONTROL MEASURES

Once the root causes of the event have been established, the investigators can propose effective risk control measures to prevent a recurrence of a similar event. Action plans should be compiled, and responsibilities allocated for these actions. All actions should have objectives that are time bound.

CONCLUSION

By following the *Logical Sequence Accident Investigation Method*, the investigator will have completed a systematic and thorough investigation of the accident. Each stage of the investigation must be completed fully before moving on to the next phase (next domino). Mistakes are made when investigators try to pre-empt the investigation conclusions and jump from phase to phase during the investigation.

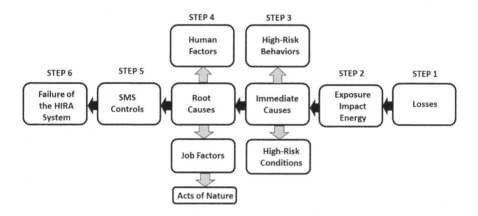

FIGURE 14.8 The *Logical Sequence Accident Investigation Method.*

15 Determining the Losses

USING THE *LOGICAL SEQUENCE ACCIDENT INVESTIGATION METHOD*

Effective accident investigations start at the end result of the event, that is, with the losses. All losses are first determined. These could be injuries, property or equipment damages, harm to the environment, business interruptions, or a combination of losses.

The impact, exposures, or energy exchanges are then identified. The proximate or immediate causes of the unplanned release of energy, in the form of high-risk behaviors and unsafe work environment, are then ascertained.

Working back from each immediate cause, a root cause analysis is done by asking "Why?" the act was committed or "Why?" the work environment was unsafe.

The weakness in the management control systems is then determined, and the failure to identify hazards and assess their risks is established. Based on this analysis, risk control measures can be proposed. This is the *Logical Sequence Accident Investigation Method.*

LOGICAL STARTING POINT

Visiting the accident site and determining and recording the consequences of the event is the logical place to begin the investigation. All accidents result in some form of loss, which is the end consequence of the accident sequence. Once the consequences of the accident, in the form of losses, have been determined, the investigator can then identify the exposures, impacts, and exchanges of energy that caused those losses. Only then should the investigation move to identify the immediate causes.

RISK ASSESSMENT

A risk assessment is a projection of what could happen during a work process, and what the results could be. Although a risk assessment cannot accurately determine if and when an event may occur, it should take into account what measures could be taken to reduce the probability of an occurrence, and the consequences, should it happen. The object of the assessment would therefore be to minimize the consequential losses of the event, by setting up suitable contingency measures and plans. Also, based on the risk of the situation being assessed, the exercise could indicate what steps should be taken to reduce the probability (likelihood) of an accident occurring.

POST-ACCIDENT RISK ASSESSMENT

After an accident, the situation should be risk assessed to determine what *could have* happened under different circumstances. *What happened* is important to investigators,

DOI: 10.1201/9781003220091-18

but *what could have happened* is also important when considering future preventative measures. A risk matrix that is included on the investigation form will enable investigators to do this assessment easily.

LOSSES

The consequences of an accident could be losses such as injury, disease or illness, property, equipment damage, business interruption, product and production loss, environmental harm, and other indirect losses and disruption.

AGENCY AND AGENCY PART

The agency and agency parts should also be identified, as these are the items that were closest to, and responsible for the transfer of energy, and subsequent loss. The agency is what transferred the energy to cause the loss. The portion of the agency that caused the harm or damage is the agency part. For example, a worker's hand was cut while he was operating an unguarded circular saw. In this case, the circular saw is the agency, and the blade is the agency part.

TYPES OF LOSSES

Accidents result in some form of loss. This could be in the form of fatalities, injuries, occupational diseases, product and equipment damage, or business interruption. A near-miss incident results in no loss. It should, however, be ranked for potential loss and probability of recurrence. Under different circumstances, it may have resulted in a loss.

The three main categories of loss are as follows:

- Direct losses
- Indirect or hidden losses
- Totally hidden losses

These are also referred to as insured and uninsured losses.

DIRECT LOSSES (INSURED LOSSES)

Direct losses such as medical treatment are normally covered by insurance. In most instances, worker's compensation insurance pays for a certain medical cost arising out of the accident.

MEDICAL

Modern workers' compensation laws provide fairly comprehensive and specific benefits to workers who suffer workplace injury or illness. Benefits vary but often include medical expenses, death benefits, lost wages, and vocational rehabilitation.

The National Safety Council (NSC) (US) publication *Injury Facts 2019* (2019b), gives the cost of medical expenses as a result of work accidents, at $35.5 billion, of the total cost of injuries and fatalities of $171 billion. They report that the average workers' compensation claim is in the region of $41,000. Work-related medically consulted injuries totaled 4.64 million in 2019.

(website 2021)

COMPENSATION

Compensation for wages lost, permanent disability, or death as a result of a work-related accident is also normally covered by insurance. This varies from country to country. The compensation paid is a minimal amount and does not compensate for totally hidden costs such as pain and suffering and long-term lifestyle changes.

Permanent Disability

Ongoing compensation may be paid out to an employee who suffered permanent disability as a result of an accident. In some cases, this is a lump sum payment, and in other cases, it is an ongoing benefit paid to compensate for earnings the injured person would have earned before the accident. As a result of the injury, the injured worker may have to do a more menial, lower paying job than he or she had at the time of the accident.

Rehabilitation

In most cases, rehabilitation of the injured worker is also paid by workers' compensation. This would enable the injured worker to return to work to a position where they could recover from the injury or disease. This position would not expose the recuperating worker to stressful work situations that may aggravate their injury or disease. These positions are termed light duty and form a part of the company's return to work program, which is an element of their SMS. Rehabilitation would include medical treatment for the physical rehabilitation of injuries or the fitting of prostheses.

INJURY CLASSIFICATIONS

The following are guidelines on the classification of work-related injuries and diseases:

- A *work injury* is any injury suffered by a person, which arises out of and in the course of their employment. (Wherever the word "injury" is used, it is often construed to include also occupational disease, illness, and work-related disability.)
- An *occupational disease* is a disease caused by environmental factors, the exposure to which is peculiar to a particular process, trade or occupation, and to which an employee is not ordinarily subjected or exposed outside of or away from such employment.

- An *occupational death* is any fatality resulting from a work injury, regardless of the time intervening between injury and death.
- *Permanent total disability* is any injury other than death which permanently and totally incapacitates an employee from following any gainful occupation, or which results in the loss of, or the complete loss of use, of any of the following in one accident: Both eyes, one eye and one hand or arm, or leg or foot, and any two of the following not on the same limb; hand, arm, foot or leg.
- *Permanent partial disability* is any injury other than death or permanent total disability, which results in the complete loss of use of any member, or part of a member of the body, or any permanent impairment of functions of the body or part thereof, regardless of any pre-existing disability of the injured member or impaired body function.
- *Temporary total disability* is any injury that does not result in death or permanent impairment but renders the injured person unable to carry on his or her normal activities of their employment (the job he or she normally does) during the entire time interval corresponding to the hours of his or her regular shift on any one or more days (including Sundays, days off, or plant shutdown) subsequent to the date of the injury. Also, all fractures and bone damage, including such fractures that do not result in permanent impairment or restriction of normal function of the injured member, result in no lost time.
- A *medical treatment injury*, or non-disabling injury, is an injury that does not result in death, permanent impairment, or temporary total disability but requires medical treatment, including first aid.
- A *disabling injury* is a work injury that results in death, permanent total disability, permanent partial disability, or temporary total disability, for longer than a work shift.

These are the injuries used in calculating the standard injury frequency and severity rates. Lost time injury is the same as disabling injury.

In both these classifications, the injured person must return to the task they were doing at the time of the accident. Returning to light duty, or a different task, means the accident consequence was so severe that the person could not carry out their normal duties, so the injury is then a disabling or lost time injury.

DAMAGE

An organization will normally experience more property damage accidents than injury-causing accidents. Damaged property calls for repairs or replacement. The removal of damaged equipment also calls for resources that cost money. These events are significant and should be investigated, as under different circumstances, workers may have been injured by the same exchange of energy that caused damage. These accidents cost the company money, and these unplanned losses should be treated as important as other accidents. They should be investigated so that the root causes can be determined and recurrences can be eliminated.

INDIRECT LOSSES (UNINSURED LOSSES)

There are often more indirect or hidden losses after an accident than there are direct losses. Indirect losses are sometimes difficult to identify and difficult to calculate their costs. Nevertheless, they are losses to the company and disrupt the normal production activity. Some experts put these costs as being three to ten times higher than the direct costs.

Many of the indirect losses experienced as a result of accidents incur costs that are uninsured. That means that in many instances they are not covered by insurance policies, and therefore the organization must bear the costs. In some cases, these costs may amount to more than the direct costs of the accident. A history of major events has shown that in many cases, organizations have been financially crippled by costs arising out of accidents. Examples are Bhopal, Piper Alpha, the Ocean Ranger oil rig disaster, and others.

FINANCIAL LOSS

Raw material, finished products or machinery, tools, and equipment could be damaged by accidents. Buildings, structures, and motorized equipment could also be damaged. These damages cost money to repair or replace equipment and materials. Organizations have normally not budgeted for these repairs or replacements, and insurance claims can result in future higher premiums.

INTERRUPTION

An accident interrupts the process at hand. It may result in production lines being stopped or machinery and tools having to be switched off. Rendering help to injured workers or containing the resultant damage causes a temporary work stoppage. Other workers are curious and leave their workstations to be of assistance, or just to see what happened. In some cases, entire work sites are shut down until the investigation is completed. This creates major business interruptions.

PRODUCTION LOSS

After an accident, there is invariably a loss of production. This could be a loss of units produced. A section of the workplace may have to be temporarily shut down. Normal processes are interrupted and this hampers production. After an accident, workers tend to group, stand around, and discuss the event, which leads to a loss of workhours.

ACCIDENT INVESTIGATION COSTS

An accident investigation costs money in the form of workhours lost during witness interviews, site inspections, replacement of damaged equipment, site repairs, and clean-up. Should a legal investigation take place, this will occupy hours of employees' and management's time during discussions and interviews by the legal investigators. The production of the investigation report takes time, as well as the investigators' time spent doing the investigation.

LEGAL ENQUIRY AND PENALTIES

In some cases, legal costs in the form of fines for contravention of health and safety laws and regulations may apply. Should there be lawsuits as a result of the accident the costs could be enormous.

INJURY FACTS 2019

The National Safety Council (NSC) (US) gives the following costs in *Injury Facts 2019* (2019c):

> The total cost of work injuries in 2019 was $171.0 billion. This figure includes wage and productivity losses of $53.9 billion, medical expenses of $35.5 billion, and administrative expenses of $59.7 billion.
>
> This total also includes employers' uninsured costs of $13.3 billion, including the value of time lost by workers, other than those with disabling injuries, who are directly or indirectly involved in injuries, and the cost of time required to investigate injuries, write up injury reports, and so forth. The total also includes damage to motor vehicles in work-related injuries of $5.0 billion and fire losses of $3.7 billion.
>
> *(website 2021)*

TOTALLY HIDDEN LOSSES

After an accident, there are other losses and costs that are seldom seen or considered, but they are nevertheless losses that would not have been incurred if the accident had not happened.

SUPERVISOR'S TIME

The supervisor may lose time in assisting the injured employee and then arranging for work continuation. The investigation report and follow-up after the accident also consume valuable time. If a temporary or new employee is hired to take the place of the injured worker, the supervisor must train this replacement how to do the work. Attending the accident enquiry meeting also takes time away from his normal duties.

PAIN AND SUFFERING

It is impossible to put a financial figure on pain and suffering experienced by injured workers after an accident. It is almost impossible to equate a human life in financial terms. Rehabilitating from a serious workplace injury takes time, pain, and suffering.

POST-ACCIDENT SHOCK

An accident is a traumatic event, for both the employees involved and the other workers. This is post-accident shock, similar to post-traumatic stress disorder (PTSD), and results in a slowdown of production, as extra care is taken to prevent any similar event happening. An acute awareness of caution is created by an accident and has an effect on employees in the area. This is a difficult loss to measure, but it is neverthe-less a loss that can impact production.

PTSD has been described as a mental health condition that's triggered by either experiencing or witnessing a terrifying event. Symptoms may include flashbacks, nightmares, and severe anxiety, as well as uncontrollable thoughts about the event.

Loss of Skill

An injured employee is away from work, and the organization therefore experiences the loss of a skilled worker. Workers may have to be replaced by temporary workers who are less skilled and experienced than the injured worker. This could lead to a further loss of production and the unplanned cost of hiring temporary labor.

Family

An accident in which an employee is injured can have a devastating effect on the injured person's family. If he or she happens to be the breadwinner, the source of income may stop until compensation is paid out. Normal day-to-day living is dis-rupted, and in some instances, lifestyles are changed permanently.

Reputation

Any organization wants to maintain a good name. Headline news about accidents and workers being killed or injured harms this reputation and is not good for business or future recruiting of workers.

Publicity

Along with catastrophic accidents come sensational but damning publicity. Fires, explosions, and other major adverse events generate a lot of publicity, which is not good for the company image and reputation. A cost cannot be put on the financial loss as a result of this adverse publicity.

Investigation Work Stoppage

Some accident investigations can be lengthy processes involving different legal agen-cies. Under some circumstances, work is halted until the investigation is completed. This interruption cuts into production time and results in financial loss to the com-pany. Although indirect and intangible, the investigator should be aware of these losses incurred.

Legal

Should the company be found to have been negligent, lengthy legal battles may result. The local safety and health administrations may sue the company for failure to abide by the regulations. Heavy fines and citations, which have not been budgeted for, may follow. The families of the victims may take legal action for compensation for fatalities or injuries sustained as a result of the accident. These legal proceedings create a financial drain on the company and also tarnish its reputation.

CONCLUSION

Workplace accidents cost money. They are interruptions to the work process and can have devastating effects on an organization. An accident is an opportunity to determine what happened and what the cost incurred were.

Effective accident investigations start at the end result of the event, with the losses. All losses must first be determined. These could be injuries, property damage, business interruption, or a combination of losses. Only once all the losses have been identified, can the investigation proceed to establish the exposures, impacts, and exchanges of energy that caused the losses.

16 Identifying the Exposures, Impacts, and Energy Exchanges

CAUSE OF LOSS

Once the losses have been established, the investigation now identifies what exposures, impacts, or exchanges of energy took place to cause the injury, damage, or interruption. This involves singling out the agencies and agency parts that were responsible for the unintended energy exchanges that caused the losses. The exchanges of energy must have been greater than the threshold limit of the body or structure to have caused injury, illness, damage, or loss.

AGENCY

The agency is the piece of equipment or object closely associated with the loss, and is the substance, object, or radiation most closely associated with the accident's consequence. The agency could be the principal objects, such as tools, machines, or equipment, involved in the accident, and is usually the object inflicting injury or property damage.

The agency is, therefore, a key factor in the exchange of energy. It is the agent that is responsible for the energy transfer to the recipient, who consequently suffers a loss in the form of injury, occupational illness, or disease. Or, in the case of a property damage accident, damage to equipment, business interruption, or both.

TWO TYPES OF AGENCIES

There are two major classifications of agencies that are involved in the exposure or exchange of energy. They are *occupational hygiene agencies* and *general agencies*.

Occupational Hygiene Agencies

• Chemicals	• Fire/Smoke	• Ergonomics
• Dust	• Gas/Fumes/Vapors	• Illumination
• Fumes/Vapors	• Radiation	• Ventilation
• Noise/Vibration	• Fungus/Mold	• Other
• Heat/Cold	• Biological	

Occupational hygiene agencies are those items that are closest to, and cause the illness or the occupational disease. An occupational disease or illness as a result of an accident is classified in the same way as an injury.

DOI: 10.1201/9781003220091-19

General Agencies

• Walkways	• Chemicals
• Ladders	• Conveyors
• Sharp edges	• Electrical apparatus
• Power tools	• Elevators
• Machinery	• Hand tools
• Equipment	• Highly flammable and hot substances
• Animals	• Hoisting apparatus
• Boilers and pressure vessels	• Power transmission equipment
• Prime movers and pumps	• Working surfaces
• Radiation and radiating substances	• Other

AGENCY PART

The agency part is that part or area of an agency that inflicted the actual illness, injury, or damage. For example, a worker was ripping planks on a circular saw. To speed up production, he removed the machine guard, thus exposing the blade. During the cutting process, he was distracted and the blade cut his finger. In this case, the agency is the circular saw, and the saw blade would be classified as the agency part.

AGENCY TRENDS

Trends can be used to establish which agency is responsible for the majority of losses as a result of contact. Trend analysis can be made by listing the agency that causes the injury, disease, or damage.

EXPOSURE, IMPACT, OR ENERGY EXCHANGE, TERMINOLOGY

For many years, occupational safety nomenclature termed the type of energy exchange as the "accident type." In fact, the correct term is "Type of energy exchange, or nature of energy exchange or exposure." For example, *contact with electrical current* describes the unintentional flow of energy that caused the injury. It describes the contact with a source of energy greater than the threshold resistance of the body or substance, which results in a loss, in this case, injury.

INJURY CLASSIFICATIONS

There are three major classifications of injuries and occupational diseases. They are as follows:

- *Acute* – An acute injury is an injury that occurs suddenly and is usually associated with trauma such as cracking a bone, tearing a muscle, or bruising. It could be as a result of falling from a height or being crushed in a machine. Acute injuries occur in a short, almost instant, time frame.
- *Chronic* – Chronic injuries are injuries and harm to the body that occur over a period of time. Most occupational diseases are termed chronic injuries because they manifest only after many hours or years of exposure to the hazard.

Musculoskeletal disorders are examples of chronic injuries. They account for approximately one-third of all workplace injuries. Examples are as follows:

- Lower back injuries
- Carpal tunnel syndrome
- Rotator cuff injuries
- Tendinitis
- Trigger finger

- *Non-impact* – Another sub-classification of injuries are non-impact injuries that result from excessive physical effort directed at an outside source. These common activities include pushing, lifting, turning, holding, throwing, or carrying. Repetitive motion injuries are caused by microtasks, and result in stress or strain to some part of the body due to the repetitive nature of the task, typically without strenuous effort, such as heavy lifting.

COMMON CLASSIFICATIONS OF INJURY-CAUSING EXPOSURES OR IMPACTS

The following are the most common main categories of injury-causing exposures or impacts:

• Struck by object or equipment	• Motorized land vehicle events
• Struck against object or equipment	• Intentional injury by other persons
• Caught in or compressed by object or equipment	• Injury by person – unintentional or intent unknown
• Fall to the lower level	• Animal and insects
• Fall on the same level	• Fires and explosions
• Slips, trips without fall	• Exposure to harmful substances
• Overexertion in lifting or lowering	• All other events or exposures
• Repetitive motion involving microtasks	

EXAMPLES OF EXPOSURES, IMPACTS, AND ENERGY EXCHANGES

- *Struck by Object or Equipment* – While walking from the office to the warehouse, the injured person was struck by a moving fork-lift truck and suffered a broken leg.
- *Struck Against Object or Equipment* – While walking down the aisle between the boiler and the work area, a worker misjudged the corner and bumped into the side of the storage rack causing bruising to his right leg. His leg struck against the storage rack.
- *Caught in or Compressed by Object or Equipment* – Trying to take a short-cut through the factory, the employee moved behind a reversing trailer that crushed him against the wall causing serious internal injuries. He was caught in between the wall and the vehicle.

- *Fall to the Lower Level* – During a maintenance task, a worker slipped from the handrail he was leaning against and fell to the floor level below. He was injured as a result of his fall to a different level.
- *Fall on the Same Level* – While changing the globes in a light fitting, the maintenance man lost his balance and fell off the ladder landing on the floor. He fell on the same level.

An example of a slip, trip *with a fall* is: While walking through the processing area, the employee was reading messages on her cell phone, and did not see the notice warning of a wet floor ahead. She slipped and fell onto the floor, injuring her elbow. The injury was caused by a fall on the same level.

- *Slips, Trips Without Fall* – Because of poor housekeeping, an employee tripped over an obstruction, almost fell, and twisted her back. Although she did not fall, the trip caused her to jerk her body that injured her back.
- *Overexertion in Lifting or Lowering* – Failing to heed the basic rules of lifting heavy objects, the employee tried to move a large box from the back of the vehicle and hurt her back. The weight was too heavy and awkward for her and she overexerted her back.
- *Repetitive Motion Involving Microtasks* – Typing on a keyboard for long periods, without frequent breaks, caused a worker to suffer from tendonitis as a result of the repetitive motion. Tendonitis is also sometimes referred to as trigger finger, or tennis elbow. Other repetitive strain injuries are carpal tunnel syndrome and bursitis, which causes weakening of movement of a limb and which can be very painful.
- *Roadway Incidents Involving Motorized Land Vehicles* – A truck driver was injured when his vehicle was involved in a multiple vehicle pileup on the highway. The pileup was due to poor visibility caused by mist. This category does not include injuries caused by water vessels. This energy exchange type was the cause of most work-related fatalities in the US during 2019.
- *Intentional Injury by Other Persons* – Tempers among employees flared at a foundry as a result of a spoilt batch of aluminum castings. One employee had operated the crane before being instructed to do so, resulting in the batch of castings being ruined. The foreman was so angry, he pulled the crane operator out of the cab, and struck him in the face with his fist.
- *Injury by Person – Unintentional or Intent Unknown* – A construction worker accidentally and without intent dropped a piece of lumber onto a fellow worker below. The timber struck the employee on the shoulder causing an abrasion. The injury was caused by the unintentional action of a fellow employee.
- *Animals and Insects* – Animals and insects can cause serious injury to workers on the job. In one case, a farm worker was attacked by a bull and suffered serious injuries. In another case, while installing an outside light, an electrician disturbed a wasps' nest and was stung by the angry wasps.
- *Fires and Explosions* – Fires and explosions cause serious and fatal injuries to employees. Fires and explosions were the sixth highest cause of work fatalities in the US during the year 2019.

- *Exposure to Harmful Substances or Environments* – Asbestos laden atmospheres, areas where toxic gasses are present, and other environments that contain harmful substances can lead to serious harm or fatal injuries to employees. Some toxic environments can be entered into for brief periods without harm. Some cannot be entered into at all without the risk of being overcome and injured. Smoke, fumes, and mists are other examples of hazardous environments. Noise zones also fall into this category, as do areas where excessive radiation is present.
- *All Other Events or Exposures* – The main categories of exposures, impacts, and energy exchange have been discussed, but there are many sub-categories of these hazardous environments. During an investigation, the accident investigator must single out the specific hazardous environment or exposure that caused the illness or injury.

IMPACT TRENDS

As part of the injury recording and analysis system at an organization, an analysis of the energy exchange types can be recorded and plotted. This will give an indication of what sort of impact is happening more frequently so that remedial measures can be directed at preventing this type of energy exchange.

For example, during 2019, the top six impact types (injury causes), in the US according to the NSC publication *Injury Facts 2019* (2019d), were as follows:

1. Overexertion, bodily reaction
2. Falls, slips, and trips
3. Contact with objects/equipment
4. Transportation incidents
5. Violence and other injuries by persons or animals
6. Exposure to harmful substances or environments (website 2021)

FATAL STATISTICS

According to the US Bureau of Labor Statistics table, *Fatal Occupational Injuries by Event*, the most work-related fatalities experienced at work during 2019 were as a result of the following exposures, impacts, or exchanges of energy:

- Transportation-related accidents
- Falls, slips, and trips
- Contact with an object and equipment
- Violence and other injuries caused by people and animals
- Exposure to harmful substances or environments
- Fires and explosions (website 2021)

CONCLUSION

Once the investigation has established all the exposures, impacts, and exchanges of energy that led to the losses of the accident, the investigation can now start determining what the immediate causes of these exposures, impacts, or exchanges of energy were.

17 Immediate Cause Analysis – High-risk Behaviors (Unsafe Acts)

INTRODUCTION

High-risk behavior (unsafe act) is a departure from a normal accepted or correct work procedure, which reduces the degree of safety of that procedure.

These acts can be described as any activity carried out by workers, which are not according to the prescribed safety standard, or practice, and which can cause accidents, or have the potential to cause accidents. Such high-risk behaviors may be due to a poor attitude of workers, a lack of awareness of safety measures, or not following job safe practices as a result of other hidden or root causes. High-risk behaviors may be as a result of human failure, errors, or violations.

A SAFETY MYTH

For many years, the unsafe act or high-risk behavior of workers has been cited as the cause of the majority of work accidents. Some philosophies state that up to 90% of accidents are caused by high-risk worker behavior. Many of these statements cannot be supported by research, and are a carryover of H.W. Heinrich's research, which stated that 88% of accidents were caused by the unsafe acts of people. This research was done in 1929 and has never been substantiated, but is still believed today.

HIGH-RISK BEHAVIOR IS NOT A NEAR-MISS INCIDENT

Numerous high-risk behaviors are committed daily but do not result in a contact with energy, or impact of any sort. These are not near-miss incidents, as there has been no flow of energy that, in a contact situation, would have caused injury, damage, or other losses. Many confuse the high-risk behavior and the high-risk workplace condition with near-miss incidents. To fall into the latter category, there must be a flow of energy.

A COMPLEX SITUATION

The term high-risk behavior or unsafe act describes a complex situation. High-risk behavior is not merely a worker blatantly defying safety rules and regulations. It is the end action that results from an accumulation of a number of breakdowns and weaknesses in a management system, which was designed to keep the worker safe at work.

DOI: 10.1201/9781003220091-20

Accident investigators should look beyond the high-risk actions uncovered by the investigation, and seek the deep-rooted causes for these actions. As the Health and Safety Executive (HSE) (UK) (1999) states:

> Many accidents are blamed on the actions or omissions of an individual who was directly involved in operational or maintenance work. This typical but short-sighted response ignores the fundamental failures which led to the accident. These are usually rooted deeper in the organization's design, management and decision-making functions.

(p.6)

HUMAN FAILURE

Human failure is a broad term often used in an accident investigation. There are two main types of human failure, *inadvertent failure* (error) and *deliberate failure* (violation).

INADVERTENT FAILURE

Inadvertent failures are classified as mistakes that could be rule-based or knowledge-based mistakes. Workers do make mistakes but they are not the prime accident cause. According to Bill Hoyle (2005) in his paper, *Fixing the Workplace, Not the Worker*,

> When you read a newspaper account of an industrial accident it will almost always conclude that the cause of the accident was worker error. In a society largely based on individualism, the idea that worker mistakes are the primary cause of accidents rings true with most people. There is no denying that workers make mistakes. However, in every industrial accident, there are almost always several management safety systems involved which may not be readily apparent.

(p.3)

DELIBERATE FAILURE

Deliberate failures are violations. These violations could be routine or could be as a result of a certain situation or exceptional circumstance.

OTHER FAILURES

EXCEPTIONAL FAILURE

Exceptional failure is where a person attempts to solve a problem in highly unusual circumstances and takes the calculated risk by breaking the rules.

ACTIVE AND LATENT FAILURES

High-risk behaviors are sometimes referred to as *active failures*. These are errors and violations that have an immediate negative result.

According to the HSE (UK) (1999),

Active failures have an immediate consequence and are usually made by frontline people such as drivers, control room staff, or machine operators. In a situation where there
is no room for error these active failures have an immediate impact on health and safety.

Latent failures are made by people whose tasks are removed in time and space from
operational activities, for example, designers, decision makers, and managers. Latent
failures are typically failures in health and safety management systems (design, implementation, or monitoring). Examples of latent failures are as follows:

- Poor design of plant and equipment
- Ineffective training
- Inadequate supervision
- Ineffective communications
- Uncertainties in roles and responsibilities

Latent failures provide as great, if not a greater potential danger to health and safety as
active failures. Latent failures are usually hidden within an organization until they are
triggered by an event likely to have serious consequences

(p.11)

ERRORS

An error is an action that fails to produce the expected result, which may produce an
undesired and unwanted outcome. Human error is commonly defined as a failure of a
planned action to achieve the desired result. There are four main categories of error:

SLIPS OR LAPSES

Slips or lapses are unplanned actions. They are unintended actions that sometimes
occur when the wrong step is taken, or due to a lapse, a step of a procedure or process
is not done correctly or is left out.

A *slip* happens when a person is carrying out familiar tasks automatically, without thinking, and the person's action is not as planned, such as operating the wrong
switch on a control panel.

A *lapse* happens when an action is performed out of sequence, or a step in a
sequence is missed.

MISTAKES

Mistakes are made when decisions and actions are taken and later discovered to be
incorrect, although the employee, at the time, thought they were correct. They are failures in a plan of action. Even if the execution of the plan was correct, it would be
impossible to achieve the desired outcome.

Rule-based mistakes happen when a person has a set of rules about what to do in
certain situations and applies the wrong rule.

Knowledge-based mistakes happen when a person is faced with an unfamiliar
situation for which he or she has no rules, uses his or her knowledge, and works from
first principles, but comes to a wrong conclusion.

LATENT ERRORS

Latent errors are problems, or traps hidden within systems, which under certain conditions will contribute to an error occurring. They may lie dormant for some time, but given a certain set of circumstances they manifest.

VIOLATIONS

Violations are deliberate deviations from safe work standards and procedures. They can be accidental, unintentional, or deliberate. Violations are rule-breaking actions and are deliberate failure to follow the rules, such as cutting corners to save time or effort, based on the belief that the rules are too restrictive and are not enforced anyway. The different types of violations are as follows:

Routine Violations

Routine violations are identified in most violation categories. Routine violations occur when the normal way of doing the work is different from prescribed rules and procedures. Often routine violations are so common among work teams that they are no longer perceived as violations or high-risk behaviors. This is called "that's the way things are done around here."

Unintentional Violations

Unintentional violations occur when rules are written which are almost impossible to follow. This could occur when workers do not know or understand the rules that they are expected to follow. An example of an unintentional violation would be the violation of the speed limit posted in the parking lot of a warehouse which reads 5 MPH (8 km/h), which is far too low for the area and which is almost impossible to maintain. Organizations that profess an injury-free culture are at a loss for words when questioned about paper cut injuries that occur in the offices. Although they are injuries, they are not regarded as injuries in an injury-free culture. This is a clear case of an unintentional violation.

Procedural Violations

Procedural violations happen when procedures are purposefully deviated from, ignored, or bypassed. This is often summarized as "failure to follow procedures." The reason for this may be that the procedure is incorrect, outdated, or difficult to follow.

Exceptional Violations

Exceptional violations occur when an isolated departure from procedure occurs. This type of violation is neither typical of the employee nor condoned by supervision. Exceptional violations occur in unusual circumstances. In some crises, these violations may even be inevitable, especially when it is believed that the violation is necessary to cope with the exceptional circumstances.

Situational Violations

Situational violations occur when circumstances in the workplace, such as time, pressure, or a sense of urgency, require or encourage employees to violate safety rules. In the case study discussed (Accident Scenario 2), the team of electricians was working under pressure to connect a customer to the electricity supply so they climbed up, and worked on poles that had not been backfilled nor supported.

THE TOP 10 OSHA VIOLATIONS

According to *Occupational Health and Safety*, in the 2019 fiscal year, OSHA issued approximately 27,000 citations combined in the following categories which were the top 10 violations in the US:

- Fall protection – general requirements
- Hazard communication
- Scaffolding
- Lockout/tag out
- Respiratory protection
- Ladders
- Powered industrial trucks
- Fall protection – training requirements
- Machine guarding
- Personal protective equipment (PPE) – eye and face protection (OHS website 2020).

OTHER TYPES OF ERRORS

PERCEPTUAL ERRORS

Perceptual errors occur when an operator's sensory input is degraded and a decision is made based upon this faulty information. Unclear signals or mixed messages can cause perceptual errors to occur.

RULE-BASED ERRORS

Rule-based errors are when a worker applies written or memorized rules to deal with an unfamiliar situation. Rule-based errors are situations where the use, or disregard of a particular rule, or set of rules, results in an undesirable outcome. Some rules that are appropriate for use in one situation may be inappropriate in another. The misinterpretations of rules, or deviations from prescribed procedures, lead to mistakes. This may happen when changes in the situation prevent an individual from relying on skills.

SKILLS-BASED ERRORS

Skills-based performances are situations in which workers' perform a task with little conscious thought. This is usually the result of extensive experience with a given operation. These are actions people do, almost without thinking, like riding a bicycle or typing a letter.

When operating in a skill-based performance mode, most mistakes are due to inattention. These errors occur in experienced situations. They occur in the worker's execution of a routine, often well-practiced task. They could be due to a memory lapse or slip of action. In a skill-based mode, workers rely on the work experience of having done the same task with little or no attention over the years.

DECISION-BASED ERRORS

Choosing the wrong course of action may result in an unsafe situation. Mistakes are *decision-making* failures. They arise when workers do the wrong thing, believing it to be right.

KNOWLEDGE-BASED ERRORS

Knowledge-based performance relies on a worker's understanding of a task. Many errors result from flaws and weaknesses in that understanding. This is also known as knowledge-based mode or lack of knowledge-based mode. Because knowledge-based performance relies on an individual's knowledge-based performance, when workers don't know what they're doing, such as when faced with totally unfamiliar situations, they rely on the existing knowledge to help them. They look for patterns, and apply a remedy they've learned from other tasks, to the situation facing them. Sometimes, a wrong step is applied because of this lack of knowledge.

ERROR CHAIN

There is a concept in aviation called an error chain. An error chain can be defined as *A sequence of minor mistakes leading to a disaster,* or *a series or chain of events culminating in a loss.* This means it is not a single error, not two errors, but several successive errors in judgment or execution. If any of them had been avoided, there would have been no loss.

The error chain refers to the concept that many contributing factors typically lead to an accidental loss, rather than one single event. Each link in the error chain is an event that contributes to the loss. This supports the principle of multiple causes.

The error chain can be just a single link where just one mistake can end in disaster, or it can be many links where things all have to line up perfectly for the loss to happen. Breaking the error chain is when someone intervenes to stop a chain of events that, if allowed to continue, would ultimately result in an unplanned loss.

CATEGORIES OF HIGH-RISK BEHAVIORS (UNSAFE ACTS)

High-risk behaviors, also termed unsafe acts, are one of the immediate causes of the unplanned impact and exchange of energy that results in a loss. (The other category of immediate causes is the unsafe workplace conditions.) A description of the most common categories of high-risk behavior follows.

OPERATING EQUIPMENT WITHOUT AUTHORITY OR TRAINING

A maintenance electrician had switched off a bottling machine and was busy replacing the drive belts. He did not physically lock out the machine, but merely hung a notice on the switch stating "do not start." Upon arriving for his shift, the filler operator noticed that the machine was not working. Proceeding to the electrical controls he saw the sign saying that he must not start the machine but, without authority, he started the machine which in turn injured the electrician severely.

FAILURE TO WARN

Accidents are often the result of someone not issuing a warning. An employee started the fork-lift truck which was parked because the hydraulic fluid level was low. The vehicle did not have brakes because of the shortage of hydraulic fluid. Fortunately, he managed to stop the vehicle with the parking brake. Here, there was a failure to warn that the fork-lift truck was unserviceable.

The non-reporting of near-miss incidents would also be a failure to warn. Near-miss incidents are warnings that something serious may happen next time round, and are precursors to accidents. If ignored, these warnings would go unreported and unheeded. Unreported property damage accidents should also serve as a warning of some form of system failure.

FAILURE TO SECURE

A vehicle driving on the freeway was struck by a spade which flew off the back of the pickup traveling in front. Although no damage or injury resulted, this was a failure to secure the spades and other tools in the bed of the pickup.

IMPROPER LIFTING

Carrying an awkward load, carrying a load that is too heavy, or not lifting with your knees are all examples of the high-risk behavior of improper lifting. This could also apply to wrong rigging and hoisting practices.

NOT FOLLOWING PROCEDURES

Many tasks in the workplace require the following of written procedures. Some critical tasks have step-by-step procedures that are vital to follow for the safe execution of the task. Deviation from, or not following these procedures, is high-risk behavior.

WORKING AT AN UNSAFE SPEED

An apprentice was sent to fetch a screwdriver and a wrench from the tool store. To please his supervisor, he ran to the tool store. While running he slipped at the corner, fell, and bumped his head on a refuse bin. He was operating at an unsafe speed.

REMOVING SAFETY DEVICES OR MAKING SAFETY DEVICES INOPERATIVE

While working on a bench-mounted circular saw, the operator found that the guard over the saw blade hampered his progress. He switched off the machine, tied a piece of wire to the guard, and secured it to a brick lying on the floor. The machine was now totally unguarded. The operator had rendered the safety device inoperative.

USING UNSAFE OR DEFECTIVE EQUIPMENT

A worker had to change a light fitting in the workshop and used a wooden ladder that had a crack in the one upright. This is using unsafe and defective equipment. In another section, in the welding area, the welder used an electrode holder that was not correctly insulated. He was using defective equipment.

USING EQUIPMENT IMPROPERLY

In order to use an angle grinder as a tool sharpening grinder, instead of a bench grinder, the employee removed the guard and tied the angle grinder to a wooden box with pieces of wire. He proceeded to sharpen the tools on the machine. He was using the equipment improperly.

UNSAFE PLACING

A delivery firm delivered a load of heavy boxes to a factory. Not being instructed where to place these boxes or how to stack them, they offloaded them and stacked them haphazardly in a demarcated walkway. The placing of the boxes was unsafe.

TAKING UP AN UNSAFE POSITION OR UNSAFE POSITIONING

A worker decided to have a rest during his lunch break and lay down on a conveyor belt. The conveyor belt suddenly started moving and he was almost pulled into the crusher. He took up an unsafe position. An employee standing under a suspended load would be taking up an unsafe or improper position. Riding on a load being hoisted by a crane is taking up an unsafe position.

WORKING ON OR SERVICING MOVING OR DANGEROUS EQUIPMENT

While a 6 feet (2 m) long shaft was being turned on a lathe, the maintenance technician decided to oil the lathe's bearings. The procedure was that nobody was to work on or lubricate moving machinery. The technician was working on moving, dangerous equipment.

DISTRACTING, TEASING, OR HORSEPLAY

The process team was in high spirits and decided to play a practical joke on a colleague. They made a sign which said "kick me, " stuck it on the back of his overall, and laughed while other people followed the instructions pinned on his back. This led to horseplay in the workshop that could have resulted in accidental injury. Horseplay is high-risk behavior.

Failure to Wear PPE

Being in a hurry to complete a task, a worker failed to use safety eyewear when grinding a casting. Sparks from the casting shot up into his eye, causing him to lose a shift as a result of the injury. The worker failed to wear PPE.

Improper or Unsafe Positioning

Stacking or storing items and material in the incorrect places is an example of improper positioning. Items that protrude into walkways and other bumps against hazards are further examples.

Improper Loading

Loads that are too heavy, bulky, and not secured, is improper loading. Loads to be transported must always be secured. Loading areas should be well illuminated, free from surrounding hazards, and away from immediate traffic hazards. Overloading should be avoided.

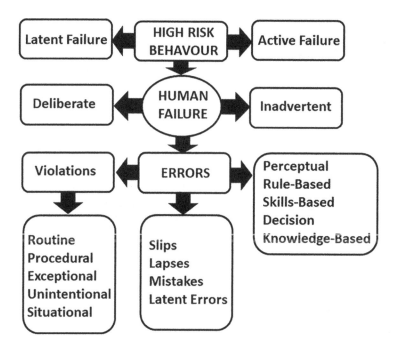

FIGURE 17.1 Human failure, types of error, and high-risk behavior.

IMMEDIATE CAUSE ANALYSIS – HIGH-RISK BEHAVIOR

During an investigation, an immediate cause analysis is done by determining what high-risk actions occurred that led to the accident. These actions could be direct or indirect actions that led to the occurrence of the event. Bearing in mind the principle of multiple causes, often more than one high-risk behavior is involved in an accident. The investigation, through a site inspection and witness interviews, must determine all the high-risk behaviors. Only once all the high-risk behaviors and conditions have been identified and recorded, can the root cause analysis begin.

18 Immediate Cause Analysis – High-risk Conditions (Unsafe Conditions)

HAZARDS

A hazard is a situation that has the potential for injury, damage to property, harm to the environment, or a combination of two or all three. A hazard is a source of potential harm. High-risk behavior and high-risk work conditions are examples of hazards.

The following are examples of the types of hazards that can be encountered in the workplace:

SAFETY HAZARDS

These are sometimes called physical hazards and are hazards that create high-risk, unsafe working conditions. For example, an unguarded machine that poses a pinch point hazard or a slippery floor which is a tripping hazard.

PHYSICAL HAZARDS

Physical hazards are hazards in the environment. These can cause injury to an employee without necessarily touching them, as these can cause injury by being exposed to them. They include the following:

- Noise
- Radiation
- Vibration
- Temperature extremes

PSYCHOSOCIAL HAZARDS

These hazards include stress, workplace violence, and sexual harassment. Psychosocial hazards can have an adverse effect on an employee's mental health or well-being. These types of hazards can cause not only psychological harm but also physical harm. These hazards could arise as a result of mental and emotional demands of the work. They could be caused by a conflict arising from job or relationship roles or ambiguity. Other causes include the following:

DOI: 10.1201/9781003220091-21

- Work fatigue
- Work overload
- Isolated work
- Changes in the workplace
- Bullying
- Customer aggression

BIOLOGICAL HAZARDS

Biological hazards can cause harm to employees. They include the following:

- Harmful dust
- Mists
- Viruses
- Bacteria
- Fungi
- Bodily fluids
- Insects
- Animals, etc.

CHEMICAL HAZARDS

Chemical hazards are hazardous substances that can cause harm. These hazards can result in both health and physical impacts. They include skin irritation, respiratory system irritation, blindness, and corrosion.

Some common chemical hazards are as follows:

- Acids
- Particulate materials
- Carcinogens
- Respiratory sensitizers
- Welding fumes
- Chemicals in liquids
- Vapors, fumes, and gases
- Silica dust
- Fiberglass fibers
- Pesticides
- Flammable liquids, etc.

ERGONOMIC HAZARDS

Ergonomic hazards are as a result of physical factors that can result in musculoskeletal and repetitive motion injuries.

Ergonomic hazards include the following:

- Awkward prolonged posture
- Improper workstations and chairs

- Frequent heavy lifting
- Frequent physical overexertion
- Excessive exposure to vibration
- Awkward repetitive movements
- Overhead work
- Pushing or pulling, etc.

HAZARD MODES

Hazards can be classified into different modes:

- A *latent* hazard is a condition that may represent a future threat.
- A *dormant* hazard is when a situation represents the possibility of a hazard to people, property, and/or the environment, but they have not yet been affected.
- An *armed* hazard is a hazard with potential harm for people, property, or the environment, which are at risk if action is not taken.
- An *active* hazard is when the hazard develops into an accident or emergency, and there has been a loss to people, property, or the environment.

HAZARD VERSUS RISK

There is often a misunderstanding of the difference between a hazard and a risk. For clarity:

A *hazard* is a source of potential harm or damage or a situation that can cause harm to a person or damage to property or the environment.

A *risk* is the chance (likelihood) of a person being injured when exposed to a hazard, and how severe the consequence may be (severity).

HAZARD RANKING

A simple hazard ranking method is the A, B, C, ranking, which is as follows:

- A-Class hazard – Likely to cause death, permanent disability, extensive property damage, or even catastrophic results.
- B-Class hazard – Likely to cause serious injury but less serious than an A class hazard, substantial property loss, or damage to the environment.
- C-Class hazard – Likely to cause minor injury, relative property damage, and minor disruption.

HIGH-RISK WORKPLACE CONDITIONS (UNSAFE CONDITIONS)

A high-risk workplace condition (unsafe condition) is any physical condition that constitutes a hazard, which has the potential to lead to an accident if not rectified. Any condition or situation (electrical, chemical, biological, physical, mechanical, and environmental) that increases the risk and possibility of accidents can be called a high-risk condition.

LATENT CONDITIONS

High-risk workplace conditions include latent conditions or failures. These are less-apparent failures in the design of processes, procedures, organizational systems, the work environment, or equipment. These are often hidden until they contribute to the occurrence of errors, or allow errors to go unrecognized until an adverse event occurs. They are sometimes referred to as accidents waiting to happen.

SAFETY MANAGEMENT SYSTEM STANDARDS

As part of the Health and Safety Management System (SMS), standards should be written for all the physical conditions in the workplace. These standards are written health and safety performance requirements that indicate responsibility and account-ability for achieving the listed actions. They are measurable management perfor-mance requirements, stipulating certain activities, and desired outcomes. Standards should be written for each element of the SMS. These would include aspects such as illumination, ventilation, housekeeping, demarcation of aisles and walkways, machine guarding, confined space entry, and others. These must be applied, communicated to stakeholders, be consistent with the health and safety policy statement, monitored, and updated as required. These standards would dictate a clean, safe, and healthy workplace, free from high-risk conditions.

They would also include processes to ensure that the workplace remains risk-free. The requirements could include hazard reporting systems, regular inspections, and six-monthly audits against the standards.

CATEGORIES OF HIGH-RISK WORKPLACE CONDITIONS

Although there are numerous high-risk workplace conditions, the following are con-sidered to be the basic categories of high-risk conditions:

UNGUARDED, ABSENCE OF REQUIRED GUARDS

A circular saw was constructed by the mechanical engineering division. It had no guard over the blade, no riving knife, or anti-kickback device. The machine was totally unguarded.

INADEQUATELY GUARDED

A compressor in the batching plant had a machine guard over the front of the v-belt drive, but the back was totally open. The pinch point was inadequately guarded.

Two in-running gears were covered by a metal mesh. The openings of the mesh were so far apart that one's hand could easily pass through the opening and into the pinch point. Although an attempt had been made to guard the machine, it was inad-equately guarded.

DEFECTIVE, ROUGH, SHARP, OR CRACKED

A wooden step ladder, after being used for many years, was discovered in the corner of a warehouse. Its rungs were cracked, broken, and in some places held together with wire. The cross brace was also cracked and wire was protruding from both sides of the ladder constituting another hazard. This ladder was defective and cracked.

UNSAFELY DESIGNED MACHINES OR TOOLS

To sharpen their chisels, workers fixed an angle grinder to a box, which was kept in position by stones packed inside the box. This was the wrong application for an angle grinder. They had just used an unsafely designed machine.

UNSAFELY ARRANGED AND POOR HOUSEKEEPING

A work crew left drums, pipes, tools, and boxes lying in the middle of a walkway. They also spilt flammable liquid over the walkway, constituting not only a trip hazard but also a fire hazard. This constituted poor housekeeping and congestion of the walkway.

CONGESTED/RESTRICTED/ OVERCROWDED WORK AREA

Congested walkways, work areas, and storage spaces are hazards and fall under this heading of congested, restricted, overcrowded areas. In one work area, goods and equipment were stored in the area where workers operated machinery. There was little room to maneuver, and no demarcated walkways were evident, meaning that movement among the congested areas was hampered and dangerous.

INADEQUATE OR EXCESS ILLUMINATION/SOURCES OF GLARE

In a work area, it was noted that during the night shift, only one weak 40 Watt bulb illuminated the area. Workers could not see the machinery clearly and had to use portable lamps when doing critical inspections. The lighting was totally inadequate and constituted poor lighting.

INADEQUATE VENTILATION

When the acid was poured into the dipping tanks, the fumes hung heavily in the air. These fumes caused the worker's eyes to burn and their noses to run. This was a case of inadequate ventilation over the tanks.

HIGH- OR LOW-TEMPERATURE EXPOSURE

Employees working in smelters or in deep underground mines are exposed to extreme temperatures. In tropical areas, employees could also be subject to high ambient temperatures and high humidity, creating a risk of heat fatigue and exhaustion.

Working outside in snow and cold conditions also poses risks to the human body. Moving in and out of cold rooms during loading procedures causes frequent temperature extreme changes, which poses a health hazard to employees.

UNSAFELY CLOTHED, NO PERSONAL PROTECTIVE EQUIPMENT (PPE), INADEQUATE OR IMPROPER PPE

In certain processes and in certain types of industry, specific PPE and clothing are required. One cannot work in a hazardous work environment wearing normal street clothes.

UNSAFE PROCESS

The process may be inherently dangerous, such as in the case of working with radioactive sources, working at heights, and working with electricity or chemicals. The unsafe process cannot always be eliminated and inherent danger is ever-present.

NOISE EXPOSURE

Experts agree that noise levels exposure to 85-dB (A) or more over a 40-hour working week could result in permanent noise-induced hearing loss. Where the ambient noise level is 85-dB (A) or more, a noise zone is created that could irreparably harm workers' hearing should they be exposed for a prolonged period. Certain impact noise levels could also cause damage to hearing acuity.

RADIATION EXPOSURE

In certain industries, employees may be exposed to sources of radiation. Exposure is monitored by means of dosimeters and exposure levels are controlled by time exposed, distance, and shielding. If these measures are not in place or are compromised, illness due to radiation exposure can occur.

An employee at an X-ray unit did not wear the protective apron for an entire shift while taking X-rays and suffered exposure to radiation.

FIRE AND EXPLOSION HAZARD

An area where explosive materials, explosive substances, flammables, or volatile vapors are present is a hazardous work environment. Certain precautions and procedures must be taken when working in these environments to prevent fire and explosions. Explosive manufacturers and petroleum refineries are examples.

INADEQUATE WARNING SYSTEM

Warning systems can range from notices and signs to sirens, bells, whistles, and reverse warning devices on heavy vehicles and fork-lift trucks, to mention a few. If these are not in operation, workers in the vicinity will not be warned of the danger, thus creating a hazardous situation. Smoke and fire detectors are examples of warning systems.

HAZARDOUS ENVIRONMENT

Inherently hazardous environments include working in areas where explosives or flammable substances are manufactured, stored, or used. Working in a confined space or underground is working in a hazardous environment, as is working at heights or underwater operations. This would include exposure to the following:

* Oxygen deficiency
* Electricity
* Temperature extremes
* Radiation
* Noise
* Traumatic or stressful events
* Air and water pressure changes
* Harmful substances

CONCLUSION

All the high-risk workplace conditions should be noted during the accident site inspection. In some instances, the condition may not be a direct immediate cause of the event but may indicate deviations from standards and procedures, indicating a lack of safety control in other areas. For example, poor housekeeping in the accident area may not have contributed to the accident, but indicated a breakdown in a standard of the SMS. If safety standards are not maintained for one or more elements of the SMS, there may be other safety standards that do not comply as well. Similar breakdowns in other safety system controls may have contributed to the accident.

Only once all the high-risk behaviors and conditions have been identified and recorded, can the root cause analysis begin.

19 Root Cause Analysis – Personal (Human) Factors

INTRODUCTION

A root cause analysis is a systematic method of determining the underlying (root) causes behind the obvious immediate accident causes of high-risk behavior and high-risk conditions. It establishes the reason why the act was committed and establishes why the high-risk condition existed. Accident root causes have also been called "the story behind the story."

PRECURSORS TO IMMEDIATE CAUSES OF ACCIDENTS

Root causes are the reasons that high-risk behaviors occur and high-risk workplace conditions exist or are created. Root causes are the fundamental, underlying, system-related reasons why an accident occurs. They identify one or more correctable system failures.

Only once root causes are identified, can meaningful management controls be put in place. Root causes are not easily recognizable, and a great deal of effort is needed to complete an effective root cause analysis.

ROOT CAUSES OF ACCIDENTS

There are three categories of root causes:

- *Personal* (human or individual) *factors*, which relate to factors surrounding the worker's actions.
- *Job* (organizational or workplace) *factors*, which relate to the organizational systems and workplace.
- *Natural causes* (acts of nature) such as hurricanes, floods, earthquakes, and other events beyond our normal control.

The terms *Personal* and *Job Factors* will be used for brevity from now on. *Natural causes* will not be expanded on.

ROOT CAUSE ANALYSIS

Root cause analysis is a structured questioning process that enables investigators to recognize and discuss the underlying beliefs, practices, and processes that result in accidents. Root causes are basic accident causal factors, which if corrected or removed may prevent the recurrence of an accident.

DOI: 10.1201/9781003220091-22

Once the immediate causes are established, the root causes will indicate where the organization failed to identify the risks and put suitable health and safety systems in place. Implementation of applicable risk reduction controls to treat the root causes will prevent similar accidents occurring in the future. In the event of a high-potential near-miss incident, the exact same technique is followed.

THE 5-WHY METHOD OF ROOT CAUSE ANALYSIS

This is a simple but effective method of root cause analysis (problem solving) and involves asking the question "Why?" for each high-risk behavior and high-risk condition identified, at least five times, until a deep-seated reason for the failure is found.

In some cases, the answer to a "Why?" may be another immediate cause in the form of a high-risk behavior or condition. The investigator should then continue asking "Why?" for each immediate cause that surfaces, until all the root causes are found.

Evaluating and criticizing the answers before the process is completed will interrupt the questioning process, and serve no purpose. The duplicated answers are a necessary part of the process of deriving the root causes.

DUPLICATION OF ANSWERS

In many instances, the answers to the "Why?" questions may seem to be duplicated or very similar. This does occur during brainstorming sessions. The investigator should however persevere and continue asking "Why?" without adjudicating or evaluating the answers, until all the responses have been obtained.

GRAY AREAS

Often high-risk behaviors are closely intertwined with high-risk workplace conditions. Sometimes, high-risk behaviors create high-risk conditions, and high-risk conditions cause high-risk behaviors, and it is difficult to separate them as they are so closely interlinked.

There is often a gray area between personal and job factors, and it is sometimes difficult to draw a hard and fast dividing line between personal and job factors as they are often intertwined. Despite this, as long as all the root causes are found and listed, then the root cause analysis will be successful.

REMEDIAL MEASURES

Only once all the root causes have been identified, can failures and inadequacies in the system be determined and rectified. Some proposed risk control solutions can be implemented within a short time period, others, such as retraining, may take longer.

AN EXAMPLE OF A ROOT CAUSE ANALYSIS

A supervisor instructed an employee to move a fork-lift truck into the loading bay. The employee was not a qualified fork-lift truck operator, but the supervisor got angry and insisted. While moving the vehicle, one fork caught the side of the metal cladding on the building and ripped a piece of the metal off.

High-risk behavior – One high-risk behavior was that the employee was doing unauthorized work, and another high-risk behavior was the conflicting instruction issued by the supervisor.

The root cause analysis is as follows:

Why? – The employee had not been trained or licensed to operate a fork-lift truck (Root Cause – Job Factor).

Why? – It was not his regular job (Root Cause – Job Factor).

Why? – He was instructed to move the fork-lift truck by the supervisor (Root Cause – Job Factor).

Why? – The supervisor was under pressure to get the job done (Root Cause – Job Factor).

Why? – The supervisor issued an unsafe instruction (Root Cause – Job Factor).

Why? – Inadequate leadership displayed by the supervisor (Root Cause – Job Factor).

High-risk condition – The high-risk condition was a vehicle being operated by an unskilled employee.

The root cause analysis is as follows:

Why? – The employee was not qualified to drive the vehicle (Root Cause – Personal Factor).

Why? – There was no other driver available (Root Cause – Job Factor).

Why? – The employee was instructed to drive the fork-lift truck by the supervisor (Root Cause – Job Factor).

Why? – Supervisory violation – rules willfully disregarded by supervisor (Root Cause – Job Factor).

PERSONAL (HUMAN) FACTORS

Personal (human) *factors* relate to the employee. They include items such as the worker's physical ability, physical size and strength, knowledge, skills, and experience. Other factors include physical or mental fatigue, stress, anxiety, depression, employee morale, and alcohol or drug abuse. Personal factors also encompass tired staff, workers who are bored or disheartened, and individual medical problems.

The main categories of personal factors are as follows:

- Inadequate physical capability
- Inadequate mental (cognitive) capability
- Lack of knowledge
- Lack of skill
- Stress, depression, and anxiety
- Improper motivation

INADEQUATE PHYSICAL CAPABILITY

The employee may be either physically or mentally unsuited for the job in hand. This may have contributed to the immediate cause that led to the accident. Physical shortcomings for the job could be a root cause of an accident.

This root cause indicates that the worker is in some way physically incapable to do the task. This could create a high-risk condition or high-risk behavior that could lead to an accident.

EXAMPLE – INADEQUATE PHYSICAL CAPABILITY

An employee had to climb up a vertical ladder attached to a tank to take the pressure readings on a gauge at the top of the tank every four hours. If the pressure went over the red line, he was to contact the control room and warn them. During one shift, there was an over-pressure warning alarm in the control room, and the plant had to shut down one boiler to prevent an explosion. No warning had been received from the employee. Upon investigation, it was found that the employee had a bad knee and had difficulty in climbing the ladder every four hours as required. He subsequently failed to read the gauge, and this led to a situation where he did not warn the control room of a high-pressure reading. The control room's over-pressure warning alarm activated and one boiler had to be shut down. He was physically incapable of doing the task.

Why was there a boiler shut down? The gauge had not been read, and therefore no warning was issued to the control room. Immediate causes were a failure to follow the procedure and a failure to warn the control room. A high-risk condition was the gauge that was mounted on top of the tank.

High-risk behavior – The high-risk behavior was not reading the gauge and not warning the control room.

The root cause analysis is as follows:

Why? – The procedure for reading the gauge was not followed (Root Cause – Personal Factor).

Why? – The gauge was not read every four hours as an employee could not climb the ladder (Root Cause – Personal Factor).

Why? – No warning was sent to the control room (Root Cause – Personal Factor).

Why? – The employee could not climb the ladder and read the gauge (Root Cause – Personal Factor).

Why? – The employee had a weak knee due to past sport injuries (Root Cause – Personal Factor).

High-risk condition – The high-risk workplace condition was the incorrect positioning of the gauge and reliance on a worker to issue warning to the control room.

The root cause analysis is as follows:

Why? – The gauge was incorrectly placed (Root Cause – Job Factor).

Why? – It was installed there by the contractors (Root Cause – Job Factor).

Why? – The position of the gauge was never specified in the design specifications (Root Cause – Job Factor).

Why? – Design specifications did not include ergonomic factors (Root Cause – Job Factor).

Why? – The company did not have a standard on ergonomics in the workplace (Root Cause – Job Factor).

Why? – There were no ergonomic inspections that would have shown that the gauge could be mounted at the eye level (Root Cause – Job Factor).

Why? – The employee had physical difficulty in climbing the ladder (Root Cause – Personal Factor).

Why? – His knee was painful and sore (Root cause – Personal Factor).

Why? – He suffered from knee pain as a result of previous sport injuries (Root Cause – Personal Factor).

Why? – He was not required to undergo a hiring vocational adaptability test upon being hired (physical examination) (Root Cause – Job Factor).

Why? – This was not deemed necessary by the company (Root Cause – Job Factor).

Why? – There were no employee job specifications for this type of work (Root Cause – Job Factor).

Why? – There was no company standard or requirement for worker job specifications (Root Cause – Job Factor).

One of the root causes of the high-risk behavior of not reading the gauge every four hours as required was as a result of the employee's physical inability to climb the ladder frequently due to weak and painful knees as a result of previous sport injuries.

This accident's root causes were as a result of a pre-employment physical, job requirement specification not being available, and a physically incapable person being placed in the wrong job. Further analysis asked why he had to climb the ladder to reach the gauge in the first place. This highlighted another root cause, in that the gauge could easily have been rerouted to an eye-level position, where it would have been read much easier, with little physical effort. A lack of ergonomic studies meant that the gauge was never identified as being in an inaccessible place, and could have been moved easily. No new plant or equipment specifications referred to the positioning of critical gauges or ergonomic considerations.

INADEQUATE MENTAL (COGNITIVE) CAPABILITY

Before employees are placed in job positions, they should undergo a complete pre-employment physical examination as well as a mental ability test. These examinations will ensure that the right person is matched for the specific job. Employees may have phobias or fears, or be suffering from some emotional disturbance, or mental illness. The task may be above the worker's intelligence level, or there is an inability to comprehend or learn instructions issued. Some adverse events may be as a result of poor judgment, poor coordination, or a slow reaction time.

Example – Inadequate Mental (Cognitive) Capability

A crane operator continuously placed loads in the incorrect position and had to reposition them on numerous occasions. On one occasion, the load was swung incorrectly and knocked into a nearby building causing damage to the structure and the load. It was apparent that machinery was being operated by an unskilled operator.

High-risk behavior – The high-risk behavior was operating equipment unsafely. The root cause analysis is as follows:

Why? – She had not been tested for reaction times during the pre-employment selection process (Root Cause – Job Factor).

Why? – The company did not see the risk and therefore no testing was done (Root Cause – Job Factor).

Why? – During her training, these weaknesses had not been identified (Root Cause – Job Factor).

Why? – Physical demand of the job not explained during hiring (Root Cause – Job Factor).

High-risk condition – The high-risk condition was a crane being operated by an employee who did not have the correct cognitive attributes for a crane driver. The root cause analysis is as follows:

Why? – The employee's reaction time was slower than needed for a crane operator as she could not operate the controls quick enough (Root Cause – Personal Factor).

Why? – She had a poor distance judging ability (Root Cause – Personal Factor).

Why? – She had not been tested for reaction times during the selection process (Root Cause – Job Factor).

Why? – There was no company policy for this (Root Cause – Job Factor).

Why? – The hiring procedure was flawed (Root Cause – Job Factor).

Why? – There was an urgent need for the crane to be put into operation (Root Cause – Job Factor).

A further question concerning cognitive abilities could be "Is the task boring for the operator?" Perhaps, the task has been done so many times it is done almost automatically. This happens often with repetitive work. The task may also lack the challenge required to keep the employee's attention or the attention ability of the person has a weakness (Figure 19.1).

Fixing the Worker

While many accident investigation results are aimed at "fixing the worker," they fail to identify the drivers and movers behind the worker's actions. In some cases, deep-seated phobias or fears could cause the worker to make mistakes. Sometimes, a person with the wrong aptitude is put into a job for which they are not suited.

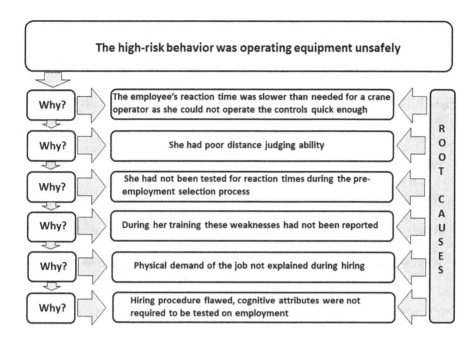

FIGURE 19.1 Deriving the root cause using the "Why?" method.

LACK OF KNOWLEDGE

A lack of knowledge is often found as a root cause of many accidents. This means that the employee committed an error, but due to a lack of knowledge, he did not recognize it as a mistake. This could be a lack of knowledge of an aspect of his or her trade or profession, or of the work they carry out, or of the safety requirements of the job. Properly directed training programs can help eliminate these deficiencies.

Example – Lack of Knowledge

A new employee was assisting a welder to chip welding slag off of welds along the side of a newly built harvester. He was chipping the slag with a hammer and small fragments of slag were flying all over. He was not wearing any eye protection. The new employee stated that he did not know he was supposed to wear eye protection when chipping. This would prompt the following questions, was the lack of knowledge on how to perform the task safely a factor?

High-risk behavior – The high-risk behavior was a failure to wear protective equipment, specifically, eye protection.

The root cause analysis is as follows:

Why? – Employee was not aware that he should wear eye protection when chipping (Root Cause – Personal Factor).
Why? – The employee did not attend health and safety orientation training (Root Cause – Job Factor).

Why? – The qualified welder did not provide the correct on-the-job training (Root Cause – Job Factor).

Why? – There was no call for a health and safety orientation review, so it was outdated and poorly applied (Root Cause – Job Factor).

High-risk condition – Employee not correctly trained in health and safety rules and protocols.

The root cause analysis is as follows:

Why? – Health and safety orientation is only presented weekly (Root Cause – Job Factor).

Why? – There are not sufficient new hires to warrant a daily orientation (Root Cause – Job Factor).

Why? – The on-the-job training did not include that specific aspect of the job (Root Cause – Job Factor).

Why? – There are gaps in the on-site training method (Root Cause – Job Factor).

Why? – It has not been revised or updated for years (Root Cause – Job Factor).

Other questions to be asked concerning training include the following:

- Was the training given based on a risk assessment?
- Were all the hazards discussed in detail?
- Was training in the specific procedure done?
- Was there a follow-up to the training?
- Was the employee's performance monitored or observed?
- Did the employee know about the safety aspects of the task?
- How long ago was the training attended?
- Did the employee receive updated training related to the task through safety briefing, job coaching, or safety practices?

HEALTH AND SAFETY TRAINING

Training in the philosophy of health and safety at work is important. It helps employees understand the cause of accidents, the effects, and the results. Training employees to understand company health and safety standards is another vital aspect of safety training. This training program could include teaching employees the safety rules of the organization, as well as general health and safety.

The organization should conduct a health and safety needs analysis to ascertain which health and safety training needs to be presented. This needs analysis should indicate who should attend the training and how often. Line management should be included in this analysis.

NEW HIRE SAFETY ORIENTATION (INDUCTION) TRAINING

Organizations should have a basic safety orientation training session for all new hires as well as for contractors and long-term site visitors. This induction should include the basic site safety rules and the do's and don'ts of the site. This induction varies in industry from a one-hour session to a full eight-hour lesson. Most mines spend at least two days on safety induction, which normally includes a day on first aid training.

On-the-Job Training

After induction training, the new employee proceeds to the workplace where on-the-job training is given. This training familiarizes the employee with the work site, the work processes, and includes the health and safety requirements of the job. Only once the employee is thoroughly familiar with these requirements, should they be allowed to work on their own.

Refresher training is normally a follow-up and can be given on an annual basis, or when a person returns to the workplace after a leave of absence.

Specific Health and Safety Training

Depending on the industry, specific health and safety training could cover topics such as

- First Aid
- Accident/near-miss incident investigation
- Modern safety management
- Safety for supervisors
- Hazard communication (HazCom)
- Regulatory safety requirements
- Firefighting
- Critical task
- Technical safety

First Aid and Emergency Training

First aid training and training in CPR (Cardiovascular Pulmonary Resuscitation) make trainees aware of their safety responsibility within a workplace, and also prepare them for emergencies. Rescue and evacuation training, as well as firefighting courses, are beneficial to both the individual and the organization.

Hazard Communication Training

Hazard communication, also known as HazCom, is a set of processes and procedures that employers and importers must implement in the workplace to effectively communicate hazards associated with chemicals during handling, shipping, and any form of exposure.

Critical Task Training

An important aspect in carrying out critical or hazardous tasks is the training that is given to enable the employee to understand the critical task procedure. Retraining after a job observation is the best method to ensure that employees follow the critical task procedures.

Technical Safety Training

A lot of safety training is technically orientated and this could include training on

- Vehicle safety
- Permit issuance and receiving
- Confined space entry
- Ladder safety
- Occupational hygiene
- Pressure vessel safety
- Asbestos awareness
- Trenching and shoring
- Lifting gear safety
- Energy control (lock-out)
- Material Safety Data Sheets, etc.

LACK OF SKILL

A skill is when a person has the knowledge, ability, and competence to complete a task within a certain amount of time, without mishap. A lack of skill is when a worker does not have the ability to complete a task, or segment of a task, successfully, or without creating a hazardous situation.

Lack of skill as a root cause of an accident can easily be demonstrated by looking at vehicle operators when they first begin to drive. They might have all the theoretical knowledge of how to drive a vehicle, but lack the experience. That is why vehicle operators need so many training hours with an experienced instructor before they have the necessary skills to operate on their own.

Example – Lack of Skill

A new worker who had recently qualified as a fork-lift truck operator was trying to stack loaded pallets onto a high storage rack. While doing so, he inadvertently knocked against an upright of the rack and bent it. He had not been instructed in how to load pallets onto high racks, nor had he practiced this maneuver.

High-risk behavior – The high-risk behavior was attempting to do a job for which he was not skilled and qualified for.

The root cause analysis is as follows:

Why? – He had not practiced this maneuver in the past (Root Cause – Personal Factor).

Why? – He attempted a lift for which he did not have the necessary skills for (Root Cause – Personal Factor).

Why? – He had just recently qualified as a fork-lift truck operator (Root Cause – Personal Factor).

Why? – He had not completed this part of the driving course (Root Cause – Personal Factor).

Why? – The operator training was scheduled for later that month (Root Cause – Job Factor).

High-risk condition – Unskilled employee operating a fork-lift truck doing high stacking maneuvers.

The root cause analysis is as follows:

Why? – Driver was doing a lift maneuver that he had not done before (Root Cause – Personal Factor).

Why? – He was not skilled in high-level loading (Root Cause – Personal Factor).

Why? – The lift was not done regularly (Root Cause – Job Factor).

Why? – Partially skilled driver was operating a fork-lift truck (Root Cause – Personal Factor).

Why? – Partially skilled driver not correctly supervised (Root Cause – Job Factor).

Not working under supervision when required, or not doing the job regularly enough, could also result in a lack of skill. A lack of experience, or a complex task, may also indicate a lack of skill. Being new to a particular job, would mean the worker has a lack of skill in performing the new and unusual task.

STRESS, DEPRESSION, AND ANXIETY

Stress is a feeling of emotional or physical tension. It can be caused by any event or thought that makes a person feel frustrated, angry, or nervous. It is a person's body reaction to a challenge or demand. Work-related stress is a harmful reaction that workers get due to undue pressures and demands placed on them at work. In short bursts, stress can be positive, such as when it helps a person avoid danger or meet a deadline. Workplace stress is the root cause of many accidents.

Other factors that can cause workplace stress, depression, and anxiety are as follows:

CHANGES

Changes to the workload, the type of work, the method of work, or the work hours can introduce stress and anxiety in workers. Changes made among the workforce can also introduce stressful conditions.

PHYSICAL CONDITION

A worker's physical condition can also lead to stress, depression, and anxiety. Being on drugs or heavy indulgence in alcohol, even prescription medicine, can lead to a changed mental state. Decreased sensory ability may occur due to temporary illness.

EXTREME EXPOSURES

Exposures to temperature extremes, environmental stresses, and variations in atmospheric pressure can lead to stress, fatigue, and lack of sleep. In many cases, personal or family problems also lead to anxiety and stress.

WORKPLACE FACTORS

Workplace factors that can lead to anxiety and stress include having to make frequent decisions, constant problem solving, and distraction by other activities, as well as confusing and conflicting demands. Boring tasks, and jobs that demand uneventful and routine monotony, are also situations that lead to stress.

According to the Health and Safety Executive (UK) (2020),

> In 2019/2020, stress, depression or anxiety accounted for 51% of all work-related ill health cases and 55% of all working days lost due to work-related ill health.

(p.3)

Example – Stress, Depression, and Anxiety

While building an overhead electric line to feed a new industrial complex, the linesmen decided to climb the poles and attach the conductors on the insulators at the top of the pole, despite the fact that the poles had not been completely backfilled or compacted. There had been a change in the method of digging the holes for the poles, and trench-like holes were dug instead of round holes, normally dug by an auger. The bulldozer that was supposed to complete the backfilling had been called away to another site. While working on the one pole, it fell. The linesman tried to jump off while the pole was falling, but unfortunately the pole fell on him and he died of his injuries.

High-risk behavior – The high-risk behavior was climbing up and working on unsecure poles.

The root cause analysis is as follows:

Why? – The wrong size holes were dug (Root Cause – Job Factor).

Why? – The wrong machine was used to dig the holes (Root Cause – Job Factor).

Why? – The linesmen had always climbed and worked on the poles before they were completely secured (Root Cause – Personal Factor).

Why? – The round holes supported the poles and they were safe to climb, the new holes were different (Root Cause – Personal Factor)

Why? – Climbing the unsupported poles was condoned past practice (Root Cause – Job Factor).

Why? – There was a sense of urgency to run the conductors and complete the job (Root Cause –Personal Factor).

Why? – After such a long delay, the linesmen wanted to get the job finished and connect the supply for the customer (Root Cause – Personal Factor).

Why? – The customer was pressurizing them to connect his electrical supply (Root Cause – Personal Factor).

Why? – There were conflicting demands (Root Cause – Personal Factor).

High-risk conditions – The high-risk conditions include the wrong machine used to dig the holes, the unsecure poles, the change in the work procedure, bad planning, and the use of incorrect machinery and methods.

The root cause analysis is as follows:

Why? – The method of digging and backfilling the holes for the poles had changed (Root Cause – Job Factor).

Why? – The auger was not available on site to dig the holes (Root Cause – Job Factor).

Why? – The job had been standing while waiting for the bulldozer which was to backfill the holes, to arrive. It arrived and was immediately deviated to another job (Root Cause – Job Factor).

Why? – Bad planning and scheduling of equipment (Root Cause – Job Factor).

Why? – They were anxious to complete the job and were under stress to run the conductors (Root Cause – Personal Factor).

Why? – The round holes supported the poles and they were safe to climb, and the new holes were different (Root Cause – Personal Factor).

Pressure from supervision and customers to complete a job can create a situation where workers take chances to complete the job. Changes to the way tasks are normally done, invariably leads to an accident, unless a change management system is in place and is applied.

IMPROPER MOTIVATION

Doing a task in a hurry or taking shortcuts are high-risk behaviors. The root causes of these behaviors are often due to motivation, but *improper* motivation. Employees want to be recognized for doing the job quickly, and this desire for recognition leads them to be motivated in the wrong way.

Example – Improper Motivation

A young employee was known to be a fast and efficient worker. He often took shortcuts during the work, but managed to finish the task in a shorter time than usual. This was a constant attempt to attract attention. His supervisor always patted him on the back and told him he had done a good job. During one shift, the task was delayed because the drill-press vice, which clamps the material being drilled on the drill press, was missing. A shaft needed drilling. To keep the job moving, he drilled the metal shaft without the vice, by holding it in place with his hand. The drill bit caught in the material and swung the shaft around cutting his left hand badly.

High-risk behavior – The high-risk behavior was using equipment incorrectly.

The root cause analysis is as follows:

Why? – The young worker was in a hurry to get the job done (Root Cause – Personal Factor).

Why? – He had a reputation for finishing the tasks at hand in a short time (Root Cause – Personal Factor).

Why? – He was recognized for his fast performance (Root Cause – Job Factor).

Why? – He was maintaining his reputation (Root Cause – Personal Factor).

Why? – He was motivated, but for the wrong reasons (Root Cause – Personal Factor).

Why? – Unsafe performance had been recognized as good performance in the past (Root Cause – Job Factor).

High-risk condition – The high-risk condition was the missing drill-press vice and an unclamped shaft in the drill press.

The root cause analysis is as follows:

Why? – The drill-press vice was not available at the drill (Root Cause – Job Factor).

Why? – Employee wanted to finish the job quickly and be recognized (Root Cause – Personal Factor).

Improper workplace performance should never be rewarded, as this leads to improper motivation and condoned practices. Workers will repeat this same unsafe behavior for recognition in the future. Good, safe performance should not be punished or ridiculed, but rather be rewarded and encouraged. Recognized good performance will be repeated. This leads to positive motivation. Inappropriate peer pressure from fellow workers also leads to improper motivation.

In many accident cases that were due to the taking of shortcuts, or rushing the job, the workers perceived there were personal benefits, such as saving time. Often this perceived benefit was to the organization's benefit and not the workers.

CONCLUSION

The acid test for any accident investigation is the finding of the root causes of the event. This can only be done once the immediate causes have first been identified. Once they are identified, a root cause analysis can be conducted. One of the most effective methods of root cause analysis is the 5-Why method, where the question "Why?" is asked at least five times, for each immediate cause. The responses will indicate the deep-seated root causes.

20 Root Cause Analysis – Job (Organizational, Engineering, or Workplace) Factors

INTRODUCTION

A root cause analysis is a systematic method of determining the underlying (root) causes behind the obvious immediate accident causes of high-risk behavior and high-risk conditions. It establishes the reason why the act was committed and establishes why the high-risk condition existed. Accident root causes have also been called "the story behind the story."

PRECURSORS TO ACCIDENT IMMEDIATE CAUSES

Root causes are the reasons why immediate causes such as high-risk behaviors occur and high-risk workplace conditions exist or are created. Root causes are the fundamental, underlying, system-related reasons why an accident occurs. They identify one or more correctable system failures.

Only once root causes are identified, can meaningful management controls be put in place. Root causes are not easily recognizable, and a great deal of effort is needed to complete an effective root cause analysis and a successful accident investigation.

ROOT CAUSES OF ACCIDENTS

There are three categories of root causes:

- *Personal* (human or individual) *factors*, which relate to factors surrounding the worker's actions.
- *Job* (organizational, engineering, or workplace) *factors*, which relate to the organizational systems and the workplace.
- *Natural causes* (acts of nature) such as hurricanes, floods, earthquakes, and other events beyond our normal control.

The terms *Personal* and *Job factors* will be used for brevity from now on. *Natural Causes* will not be discussed.

DOI: 10.1201/9781003220091-23

ROOT CAUSE ANALYSIS

Root cause analysis is a structured questioning process that enables investigators to recognize and discuss the underlying beliefs, practices, and processes that result in accidents. Root causes are basic accident causal factors, which if corrected or removed may prevent the recurrence of an accident.

Once the immediate causes are established, the root causes can be sought. They will indicate where the organization failed to identify the risks and put suitable health and safety systems in place. Implementation of applicable risk reduction controls to treat the root causes will prevent similar accidents occurring in the future. In the event of a high-potential near-miss incident, the exact same technique is followed.

THE 5-WHY METHOD OF ROOT CAUSE ANALYSIS

This is a simple but effective method of root cause analysis (problem solving) and involves asking the question "Why?" for each high-risk behavior and high-risk condition identified, at least five times, until a deep-seated reason for the failure is found.

In some cases, the answer to a "Why?" may be another immediate cause in the form of a high-risk behavior or condition. The investigator should then continue asking "Why?" for each immediate cause that surfaces, until root causes are found.

Evaluating and criticizing the answers before the process is completed will interrupt the questioning process, and serve no purpose. The duplicated answers are a necessary part of the process of deriving the root causes.

DUPLICATION OF ANSWERS

In many instances, the answers to the "Why?" questions may seem to be duplicated or very similar. This does occur during brainstorming sessions. The investigator should however persevere and continue asking "Why?" for each immediate cause that surfaces, until all the responses have been obtained.

GRAY AREAS

Often high-risk behaviors are closely intertwined with high-risk workplace conditions. Sometimes, high-risk behaviors create high-risk conditions, and high-risk conditions cause high-risk behaviors, and it is difficult to separate them as they are so closely interlinked.

There is often a gray area between personal and job factors, and it is sometimes difficult to draw a hard and fast dividing line between job and personal factors. They too are often intertwined. Despite this, as long as all the root causes are found and listed, then the root cause analysis has been successful.

REMEDIAL MEASURES

Only once all the root causes have been identified, can failures and inadequacies in the system be determined and rectified. Some proposed remedial measures can be implemented within a short time period, and others, such as retraining, may take longer.

An Example of a Root Cause Analysis

After an accident, it was found that the machine shop bench grinder was in an unsafe condition, had no fixed face shield, and the gap between the wheel and tool rest exceeded the 1/8th inch (3 mm) limit. The apprentice who fitted the wheel installed the wrong size wheel and removed the fixed face shield while he was removing the old wheel. He never replaced the shield.

High-risk condition – The high-risk condition was an unguarded machine, as the wheel and tool rest gap was too big, and the built-in face shield was missing.

The root cause analysis is as follows:

Why? – The face shield had been removed (Root Cause – Job Factor).

Why – The apprentice removed it to fit the wheel (Root Cause – Personal Factor).

Why? – It was dirty, scratched, and hampered vision of the grinding job (Root Cause – Job Factor).

Why? – Grinders were never checked or maintained (Root Cause – Job Factor).

Why? – There were no regular inspections of bench grinders (Root Cause – Job Factor).

Why? – No one was responsible for grinder maintenance and inspection (Root Cause – Job Factor).

Why? – There was no maintenance schedule for these machines (Root Cause – Job Factor).

High-risk behaviors – The face shield was not replaced, and the wrong size wheel was installed which left a gap between the wheel and tool rest that could not be adjusted.

Why? – The apprentice was inexperienced in replacing grinding wheels (Root Cause – Personal Factor)

Why? – He was working without supervision (Root Cause – Job Factor)

Why? – He forgot to replace the face shield (Root Cause – Personal Factor)

Why? – No one checked his work when he was finished mounting the wheel (Root Cause – Job Factor)

Why? – The risk of allowing inexperienced apprentices to maintain high-risk machines was not considered (Root Cause – Job Factor)

The root causes for the poor condition of the bench grinder were that the apprentice, who was unskilled in the task, was given the job of replacing the wheel. He was working without supervision and no one checked on the completed task. He did not realize the importance of the face shield and forgot to replace it when he completed the job.

There were also no pre-use inspections or maintenance schedules for these machines, which led them to be operated while in a high-risk condition. No one had been assigned the responsibility of checking the machines after maintenance. Since they were not on the monthly workshop inspection check sheet, they were overlooked.

Safety Management System Standard

A Safety Management System (SMS) standard for bench grinders should be introduced and should include the following:

- The correct wheel for the grinder
- Regular dressing of the wheel
- Fixed face shield guard condition
- The regular adjustment to ensure a 1/8th inch (3mm) gap between the tool rest and wheel
- A notice warning workers to wear protection erected at grinder
- The wearing of personal eye protection when grinding
- Regular inspections of grinders using a checklist

ROOT CAUSE ANALYSIS – JOB (ORGANIZATIONAL, ENGINEERING, OR WORKPLACE) FACTORS

One category of accident root causes is job, organizational, engineering, or workplace factors. They are related to the organization, the engineering aspects of the workplace, or the physical work environment. From here on, they will be referred to as *Job Factors*. In some examples, the *Personal Factors* are also derived as they are often intertwined with *Job Factors*.

The main categories of Job Factors are as follows:

- Inadequate leadership or supervision
- Inadequate engineering or design
- Inadequate purchasing
- Inadequate maintenance
- Inadequate tools or equipment
- Inadequate work standards
- Inadequate ergonomic design
- Wear and tear
- Abuse or misuse

INADEQUATE LEADERSHIP OR SUPERVISION

Inadequate leadership or supervision is an accident root cause that exists because an employee is managed in such a manner that the tasks are carried out in an unsafe way. Workers normally do what they are told to do, and if this direction is unclear, or ambiguous, or if high-risk behavior is rewarded, high-risk behavior becomes a condoned practice.

EXAMPLE 1 – INADEQUATE LEADERSHIP

While lifting a caterpillar track from a flatbed truck with a substitute 40-ton mobile crane, the load would not clear the flatbed. The crane operator attempted to drag the track off the flatbed, but as the load slid off the flatbed, the track flexed and swung violently, causing the crane to topple over. When the load failed to clear the flatbed, the supervisor did not give any instructions or guidance.

High-risk conditions – The high-risk conditions were using a crane, which was a substitute for the one normally used, and the urgency to get the track fitted to the machine. The root cause analysis is as follows:

Why? – The normal 50-ton crane was not available (inadequate planning) (Root Cause – Job Factor).

Why? – The track was longer than the maximum lift height of the 40-ton crane (use of incorrect equipment) (Root Cause – Job Factor).

Why? – The machine that needed the track had been idle for days, waiting for the track to be repaired and delivered (poor task planning and scheduling) (Root Cause – Job Factor).

Why? – There was a sense of urgency to unload the track (Root Cause – Personal Factor).

High-risk behaviors – The high-risk behaviors were failure to warn, dragging the load, and using inadequate equipment. The root cause analysis is as follows:

Why? –The supervisor thought the track would clear the flatbed (Root Cause – Job Factor).

Why? – Change of procedure by using the wrong size crane for the lift (Root Cause – Job Factor).

Why? – The supervisor did not stop the lift process (inadequate leadership) (Root Cause – Job Factor).

Why? – No stop signal was given by the supervisor (inadequate leadership) (Root Cause – Job Factor).

Why? – Poor pre-planning of the lift (Root Cause – Job Factor).

Why? – The crane operator did not receive any instructions once the load failed to clear the flatbed (Root Cause – Job Factor).

Why? – Inadequate instruction from the supervisor (inadequate leadership) (Root Cause – Job Factor).

Inadequate leadership could also be a result of inadequate policies or weak guidelines or a lack of procedures. Giving unclear and conflicting instructions are inadequate leadership traits. If the accident was a result of a shortage of manpower, or a shortage of funds and resources, this would be inadequate leadership as well.

EXAMPLE 2 – INADEQUATE LEADERSHIP

A scissor-lift platform could not reach high enough for workers to remove an overhead beam crane, which needed to be repaired. The platform did not extend high enough, and the workers had to climb onto the handrail to work on the crane motor. This was explained to the supervisor who insisted that the job be done and said, "I take full responsibility if anything goes wrong."

This was a high-risk act of issuing a conflicting assignment of authority. The supervisor knew there was a risk, but instructed the workers to do the job and assumed responsibility for the risk should anything go wrong.

High-risk condition – The scissor-lift could not extend high enough.

The root cause analysis is as follows:

Why? – The scissor-lift platform was not high enough for the work at hand (Root Cause – Job Factor).

Why? – There was no other platform available (poor task planning) (Root Cause – Job Factor).

Why? – Equipment was inadequate for the work at hand (Root Cause – Job Factor).

High-risk behavior – The workers had to take up unsafe work positions as the supervisor issued a conflicting assignment. Workers had to stand on the hand rail of the platform to reach the motor.

The root cause analysis is as follows:

Why? – The supervisor gave instructions and assignments that were conflicting (Root Cause – Job Factor).

Why? – The supervisor showed inadequate job planning (Root Cause – Job Factor).

Why? – He initiated and condoned the high-risk behavior (Root Cause – Job Factor).

Why? – Supervisory violation – which refers to instances when existing rules and regulations are willfully disregarded by supervisors (Root Cause – Job Factor).

Why? – Inadequate leadership to prevent accidents (Root Cause – Job Factor).

OTHER INDICATIONS OF INADEQUATE LEADERSHIP

Other indications of poor leadership are too little authority, unclear or conflicting instructions, and inadequate supervisory involvement in the SMS.

Inadequate leadership may also be a result of a weakness in the SMS, or the fact that no one has been injured doing the job in the past. Too little authority, or a lack of immediate supervision, is also an indicator of inadequate leadership.

A lack of safety commitment, conflicting instructions and objectives that conflict, and poor communication are indicators of poor leadership as well. Other questions that could be asked are as follows:

- Is there a health and safety policy statement?
- Is supervision aware of the policy contents?
- Has supervision attended general health and safety training classes?
- Are they aware of the SMS standards, policies, and procedures?

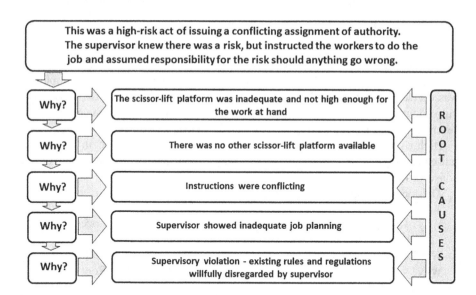

FIGURE 20.1 Deriving the root cause using the "Why?" method.

INADEQUATE ENGINEERING OR DESIGN

Inadequate engineering or design would include the inadequate design of the facility, process, line equipment, or inadequate incorporation of safety devices during fabrication. This could be caused by all the risks not being identified during the conceptual stage. A HAZOP (Hazard and operability study) not done for a new plant, process or equipment, also falls into this category.

RISK ASSESSMENTS

All risks should be considered before new equipment or new plant is commissioned or used. Hazard identification methods should be applied to ensure the design incorporates all safety features and controls. Previous accidents and accident history should be considered, and risk assessments should be applied to new processes, machinery, or products.

SPECIFICATIONS

Safety personnel should be involved in the planning and the design stage so that their input and recommendations can be incorporated. From the application of hazard identification methods and HAZOPS, specific specifications can be produced to ensure the new machinery or process lines incorporate all the safety requirements. These specifications can then be provided to contractors or suppliers and be incorporated into future purchasing or construction specifications.

EXAMPLE – INADEQUATE ENGINEERING OR DESIGN

A cement plant had contractors build an extension to their bag filling line. This modification included erecting a new conveyor belt to transport filled cement bags to the loading dock. Approximately a month after the belt was commissioned, an employee was checking the bags on the conveyor and his hand got caught in the unguarded tail-end pulley of the belt. The belt could not be stopped immediately and he suffered severe hand injuries as a result.

High-risk conditions – The high-risk conditions were found to be an unguarded tail-end pulley and the absence of an emergency stop tripwire along the length of the conveyor belt.

The root cause analysis is as follows:

Why? – Failure of the hazard identification and risk assessment process (Root Cause – Job Factor).
Why? – No routine workplace safety inspections (Root Cause – Job Factor).
Why? – Guards and trip wires were not specified in the contract (Root Cause – Job Factor).
Why? – No pre-commissioning safety inspection was conducted (Root Cause – Job Factor).
Why? – No plant specification was issued to the contractor (Root Cause – Job Factor).

High-risk behaviors – The high-risk behaviors were neglecting to design safety features into the conveyor belt, and not having a specification for the new plant which included safety considerations.

The root cause analysis is as follows:

Why? – No risk assessment was done at the conceptual stage (Root Cause – Job Factor).
Why? – No plant specification was issued to the contractor (Root Cause – Job Factor).
Why? – Legal requirements not considered during the planning stage (Root Cause – Job Factor).
Why? – Safety department was not consulted at the design stage (Root Cause – Job Factor).
Why? – No pre-commissioning safety inspection was conducted (Root Cause – Job Factor).

CHANGE MANAGEMENT SYSTEM

Other reasons for the new plant being in an unsafe condition could be that safety was not on the agenda of meetings prior to and during contract negotiations. The company SMS standards were not applied to new plant installations and not communicated to the contractor. If the organization has no change management system, changes made without company approval can lead to inadequately engineered plant modifications.

Pre-commissioning and Ongoing Safety Inspections

Safety inspections of a new plant and machinery or structures should be made on a regular basis to detect any hazards before the plant is accepted and operated. These inspections start at the design stage and continue through to the commissioning stage. All new plants, machinery, or structures must comply with local health and safety requirements, as well as local codes and regulations.

Contractor Safety Specifications

Contractors on site should follow the company requirements as set out in the SMS, and an effective contractor control system should be in place during the time that the contractors are on the site. If specifications have been issued for the new plant and equipment, deviations from these should be questioned, investigated, and rectified immediately.

EXAMPLE OF A SMS STANDARD FOR PURCHASING SPECIFICATIONS

SMS Element: Engineering Control, New Plant/Equipment and Contractor Control

Introduction

This SMS element standard will outline the controls for the purchasing and installation of a new plant, equipment, modification of present plant equipment, and the safety control of contractors.

- *New plant equipment* is described as equipment or processes added to present plant equipment.
- *Modification of present plant equipment* is the change or removal of existing plant sections.
- *Contractors* refer to direct, externally hired contractors.

Time Frame

The *Engineering Control, New Plant/Equipment and Contractor Control* standard becomes effective upon approval. This standard shall be reviewed and updated no less than annually from the date of publication, or more frequently as required.

Scope

This standard applies to all company operations. It will ensure that company business sections will understand the various safety and legal controls required when purchasing and installing a new plant, equipment, or modifying present equipment.

RESPONSIBILITY AND ACCOUNTABILITY

- Project managers are responsible for the application of this standard.
- Health and Safety Coordinators are responsible for educating project managers and other relevant persons in the application of this standard.
- Managers and supervisors are accountable for implementation and enforcement of this standard.

HEALTH AND SAFETY SPECIFICATIONS FOR PURCHASING OR MODIFYING NEW PLANT AND EQUIPMENT

- All new plant or equipment must be purchased in conformance with legal and appropriate national and international standards.
- Material and installation specifications will be drawn up for each section of plant or equipment.
- Material and installation specifications used for a new plant or new equipment must be approved by the project management department.
- Project managers are responsible for researching and recommending appropriate new material and installation specifications for a new plant or equipment.
- Where no appropriate governmental or industry standard exists, company SMS standards shall be written to apply to the equipment.
- All chemical and other hazardous materials used in the construction or modification of new, or existing plant equipment, will be covered by comprehensive material safety data sheets (MSDS) as called for in the company SMS element covering hazardous substance control.

INADEQUATE PURCHASING

Inadequate purchasing or procurement is an accident root cause. It entails whether or not the correct item was purchased, or whether it was a sub-standard item, substance, or device. Reasons for this could be that there was no research done initially, or there were no purchasing specifications drawn up, or the safety department was not involved in the purchase or procurement of the item.

HAZARDOUS SUBSTANCE CONTROL

In the case of hazardous substances, the MSDS should be requested with the purchase. Only those materials that are on an approved purchasing list should be purchased. Warning labels, operating and use instructions should always accompany the delivery of the items, as well as disposal or recycling instructions. Safety requirements such as what personal protective equipment (PPE) should be worn when working with the substance, and what first aid measures to take in an emergency, should be specified by the supplier. A safer substitute should always be sought.

SUBSTITUTE ITEMS

Substitute items, or items not on an approved purchasing specification list, always pose a risk as they call for a change in some process or procedure. Any changes should be approved by management, as well as the safety department. Workers should be informed and instructed on how to deal with these changes. Where necessary, a change management action should be instituted.

EXAMPLE – INADEQUATE PURCHASING

A consignment of 10-inch grinding wheels was purchased for a fettling department of a large foundry. Each casting had to be ground to remove sharp and rough edges. This was done by the fettling department which used more than 20 floor mounted grinding machines. While grinding a casting, the grinding wheel shattered and shot pieces of the wheel into the surrounding area. No one was injured, but when the event was investigated, it was discovered that the wheel was inferior to the wheels normally used. The material was soft and crumbly, and the circular paper label giving the safety instructions of the wheel seemed to have been poorly printed. Further examination revealed that the wheel was flawed in many respects.

The fettling department manager immediately rejected all similar wheels which were traced back to the new consignment of grinding wheels. Further investigation revealed that the wheels were obtained through a supplier that was not a company approved supplier. The country of origin of the wheels was not clear. It was later ascertained that the purchasing department had purchased the consignment because they were considerably cheaper than that from the regular supplier. Further tests showed the wheels to be inferior and hazardous.

High-risk conditions – The grinding wheels were of an inferior type, they could not take the speed for which they were rated at.

The root cause analysis is as follows:

Why? – The wheels were inferior, manufactured in some unidentified country (Root Cause – Job Factor).
Why? – The new wheels were made in a country that did not have strict manufacturing and testing codes for grinding wheels (Root Cause – Job Factor).
Why? – The price was lower than the regular supplier offered (Root Cause – Job Factor).
Why? – Inadequate purchasing procedures (Root Cause – Job Factor).

High-risk behavior – The wheels were not purchased according to the accepted standards, but purchased as a result of cost savings.

The root cause analysis is as follows:

Why? – The purchasing department did not follow the purchasing procedures (Root Cause – Personal Factor).
Why? – No pre-purchase discussions were held with the fettling department manager (Root Cause – Personal Factor).
Why? – The purchasing department wanted to save the company money (Root Cause – Job Factor).
Why? – Purchasing procedures not updated, discussed, or followed (Root Cause – Job factor).

INADEQUATE MAINTENANCE

Factors to consider if the accident root cause was a result of inadequate maintenance are as follows:

- No planned maintenance schedule
- Maintenance not done in accordance with the schedule
- Inadequate time for maintenance
- Maintenance not given priority
- Inadequate control over maintenance

EXAMPLE – INADEQUATE MAINTENANCE

An automated material elevator (skip) in an industrial plant was used to convey raw products up to the hopper which was three stories above the ground level. On the morning of January 12th, the elevator was moving the material to the top of the hopper when the automated system failed to stop its travel and it continued upward, crashing into the top beam, known as the "crash beam." The rope overwound, jamming the skip into the top of the winding wheel. This caused the rope to snap and fall, injuring two workers on the ground. The other part of the wire rope whiplashed, and tore gaping holes in the roof of the hoist house. The skip was damaged and wedged into the drive mechanism.

Investigation

The investigation showed the instruments in the control room that monitor travel positions had indicated that the skip was halfway up the shaft at the time of the collision. Further investigation showed that a physical verification of the calibration had not been done as per the planned maintenance schedule. This involved physically counting the coils of wire on the drum to verify the position of the skip in relation to the instrument reading. Previous problems had prompted the ordering of different position sensor switches, and investigators found that these had been on order for 8 months.

Witnesses Reports

The day before the accident, the control computer had been rebooted, as its numbers "had become scrambled." Enquiries indicated that this was often caused by excessive heat. On looking further, it was found that the air conditioner that cooled the control computers for all skips was not functioning correctly. Investigators found the filters of the air conditioner clogged with dirt which had reduced the efficiency of the unit, causing the computer to heat up and lose the skip stop memory. The calibration "had kicked out" as one operator put it. This, in turn, rendered the proximity warning devices useless.

High-risk conditions – The high-risk conditions were excessive computer room heat, and an absence of sensor switches, which were still on order.

The root cause analysis is as follows:

Why? – Control computer overheating (Root Cause – Job Factor).
Why? – Environment not cooled sufficiently (Root Cause – Job Factor).

Why? – Filters of air conditioner not cleaned as per schedule (Root Cause – Personal Factor).

Why? – New position sensor switches not obtained nor installed (Root Cause – Personal Factor).

Why? – Risk of an inadequately maintained sensor system high and not considered (Root Cause – Job Factor).

High-risk behaviors – The high-risk behaviors included a failure to warn, a failure to follow maintenance procedures, and not doing a physical rope coil count.

The root cause analysis is as follows:

Why? – No rope coil count had been done as per maintenance schedule (Root Cause – Personal factor).

Why? – Correlation between skip position and instrument indication not done (Root Cause – Personal factor).

Why? – Risk of skipped maintenance tasks underestimated (Root Cause – Job factor).

Why? – The maintenance plan had not been followed (Root Cause – Personal factor).

Why? – Inadequate control over maintenance (Root Cause – Job factor).

Planned Maintenance

One of the root causes was a failure to carry out basic maintenance tasks as per the planned maintenance schedule. If the maintenance schedule had been followed, the accident would not have occurred.

Inadequate maintenance also includes premature failures, faulty replacement parts, or maintenance not done according to the recommended schedule. Unclear instructions relating to maintenance and not putting maintenance as a priority are also root causes, as are maintenance tasks performed by unqualified persons, and tasks not inspected after completion.

INADEQUATE TOOLS OR EQUIPMENT

Were the tools or equipment inadequate for the job at hand? Were the correct tools available, were the wrong tool supplied, or were they in a poor condition? These are some questions to be asked to determine if the high-risk behavior or condition was a result of the root cause, inadequate tools, or equipment.

EXAMPLE – INADEQUATE TOOLS OR EQUIPMENT

The previously mentioned accident scenario under the heading "Inadequate Leadership or Supervision," which involved the rollover of a mobile crane during a lifting operation, is a classic example of the root cause – inadequate equipment. The crane was lifting a tank track (Caterpillar track) from a flatbed truck. When the crane reached its maximum lift height, the track was not lifted completely off of the flatbed. Some links were still on the flatbed. In an effort to get the track off the flatbed, the crane operator dragged the track with the crane, hoping it would slide off. Unfortunately, the track whipped in a backlash as it cleared the flatbed and this swinging motion caused the crane to topple over.

High-risk condition – The high-risk condition was using a substitute, smaller crane to the one normally used for the lift.

The root cause analysis is as follows:

Why? – The crane could not lift high enough to lift the track free from the flatbed (Root Cause – Job Factor).

Why? – The crane was a 40-ton crane when it should have been a 50-ton crane (Root Cause – Job Factor).

Why? – It had not been used for this job before (Root Cause – Job Factor).

Why? – The 50-ton crane normally used for this task was not available (Root Cause – Job Factor).

Why? – The 50-ton crane was out on loan (Root Cause – Job Factor).

Why? – The wrong mobile crane was used and it was inadequate (Root Cause – Job Factor).

One of the job factor root causes of the accident was the use of inadequate equipment, in that the wrong crane was used for the lift.

High-risk behavior – The high-risk behavior was dragging the load in an attempt to clear the flatbed.

The root cause analysis is as follows:

Why? – Because the load would not clear the flatbed (Root Cause – Job Factor).

Why? – The crane could lift no higher as it was the wrong size crane (Root Cause – Job Factor).

Why? – The job needed completion (Root Cause – Personal Factor).

Why? – The task had never been tackled with this smaller crane before (Root Cause – Job Factor).

Why? – The crane operator received no instruction from the supervisor when the load did not lift off the flatbed (Root Cause – Job Factor).

OTHER ROOT CAUSES

Equipment or tools that are dangerous, unsafe, or incorrect for the job are root causes. Defects in equipment should be reported and rectified before the equipment is allowed to be used. The correct equipment and tools should be used for the task, and any preventative maintenance specified for them should be done as prescribed by the manufacturer.

Adequate numbers of correct tools and parts, in good condition, should always be available for the task. Defective equipment should be shut down, locked out, and tagged, or marked to prevent its use until it is repaired or replaced. Equipment defects should be reported to supervision. Any changes in the normal way of doing a task should be processed through the change management program to identify and mitigate any new risks arising from the change.

INADEQUATE WORK STANDARDS

When considering inadequate work standards as a root cause, the following should be considered; were inadequate methods, procedures, practices, or rules in place at the time of the accident?

JOB SAFETY ANALYSIS (JSA)

Standards for work should be developed, communicated, and maintained. A process of identifying high-risk tasks should be carried out by a critical task identification process, and once critical tasks have been identified, procedures for these tasks should be written. This is sometimes referred to as a JSA process. Critical tasks should be listed on a critical task inventory and should be updated as needed.

TRAINING

Training in the standards, procedures, and rules should be included in the general health and safety induction training, where applicable. Ongoing training in the standards, procedures, and rules should take place on a regular basis. On-the-job training in critical tasks should also be done at regular intervals.

Regular planned task observations of critical tasks will identify weaknesses in the procedure or the execution of the procedure. The observations will also indicate the effectiveness of the training given. These procedures should be updated regularly, at least annually, or when required by a change or modification in the process that affects the procedure.

Notices and signs should be displayed to reinforce the rules of the standard or procedure. Rules should not conflict, or cause confusion among workers.

PROCEDURES

"Failure to follow procedure" is often cited as an accident immediate cause, but the investigator should always ask why the procedure was not followed. In many cases, the procedures are incorrect or hard to follow. Sometimes, they have not been updated and in many cases the training in the procedure is flawed.

EXAMPLE – INADEQUATE WORK STANDARDS

At a large industrial complex that involved smelting and casting operations as well as electricity generation, it was found that all 3,000 employees in the plant were wearing bump caps as head protection. Bump caps are not a protection for falling objects and are only intended for use where workers are in small or confined spaces, and where workers may bump their heads. Proper hard hats are worn to protect against objects and tools, which may fall from a height and cause head injury.

Bump Caps Look Better

A risk assessment found that this industry definitely called for the wearing of hard hats and not bump caps. When questioned, it was discovered that when the company had to select head protection, one general manager did not like the way he looked in a hard hat, and found he looked "more handsome" in a bump cap. So bump caps were purchased and worn by all in the plant. This continued until an outside safety consultant questioned their use of bump caps. After much persuasion, the general management changed over to more suitable head protection by purchasing and issuing correct hard hats.

High-risk condition – Inadequate PPE (bump caps) being worn by employees. Employees were wearing inadequate head protection.

The root cause analysis is as follows:

Why? – Bump caps were lighter and more comfortable (Root Cause – Job Factor).

Why? – One general manager had decided he looked better in a bump cap (Root Cause – Personal Factor).

Why? – An informal procedure was created by one person (Root Cause – Personal Factor).

Why? – No written standard for head protection (Root Cause – Job Factor).

Why? – Risk assessment not done before the decision was taken (Root Cause – Job Factor).

Why? – Inadequate work standard for PPE (Root Cause – Job Factor).

High-risk behavior – The high-risk behavior was the general manager's decision to purchase bump caps for all employees.

The root cause analysis is as follows:

Why? – The decision was not based on the risks of the workplace (Root Cause – Job Factor).

Why? – No PPE risk assessment done (Root Cause – Job Factor).

Why? – Purchasing standards inadequate (Root Cause – Job Factor).

Why? – Inadequate decision-making based on personal preference and not risk (Root Cause – Personal Factor).

Why? – Failure to warn the senior management of the risk of inadequate PPE (Root Cause – Personal factor).

INADEQUATE ERGONOMIC DESIGN

Inadequate ergonomic design refers to the way people interrelate with the work area. Ergonomics is the general term used to identify the field of activity aimed at the matching of machines, equipment and the environment, to humans, in such a way that maximum performance can be achieved. The main goal of ergonomics is that of the reduction of errors and injuries caused by repetitive or awkward motions. Ergonomics is about adjusting the job to fit the human body's needs, to improve the efficiency of the worker in the workplace, and to reduce work fatigue. In recent years, nearly a quarter of all work injuries in the US were ergonomic-related.

THREE MAIN AREAS

Ergonomics covers three main areas, anatomy, physiology, and psychology.

- *Anatomy* is concerned with the dimensions of the human body and its variations, and also includes *bio-mechanics.*
- *Physiology* includes both work physiology and environmental physiology. Environmental physiology is the effects of the physical environment on the workplace, and work physiology concerns the expenditure of energy.
- *Psychology* includes skill psychology and occupational (workplace) psychology. Skill psychology is concerned with the mental activity of information receiving, processing, and decision taking. Humans interact with machines by reading gauges and controls. They process this information and give the machine further instructions. Psychological aspects of ergonomics include the worker satisfaction, and the comfort that the person feels in their job.

Anatomy

Anatomy concerns the physical facilities such as seating, height of controls, body posture, muscular strength, and body movement. People have different body sizes and shapes and ergonomics endeavors to accommodate the average person. The belief is that the work environment should fit people, and that people should not fit the environment.

Bio-mechanics

Bio-mechanics is to do with the forces that can be applied by a human body under certain circumstances. The main anatomical considerations in ergonomics are as follows:

- Accessibility of valves, switches, and controls
- Size, shape, and comfort of seats and backrests
- Height and angle of tables and working countertops

An ergonomic checklist for seats would include the following:

- Availability of seats
- Condition of seats
- Backrest and lumbar support
- Height and tilt
- Foot and arm support
- General comfort and adjustment

A checklist for the accessibility of instruments and other equipment would include the following:

- Valves
- Instruments
- Switches
- Emergency controls

- Sample points
- Fixed ladders
- Platforms

Physiological Ergonomics

Physiology concerns itself with the effects of the physical environment on the workplace as well as the expenditure of energy.

Various factors such as illumination, ventilation, and noise affect people in different ways. Ergonomics investigates and identifies the correct temperatures for work areas, and also identifies where workers are exposed to vibration and other hazards. Work methods are studied in an effort to help prevent fatigue caused by monotonous or repetitive tasks. Tasks that are too easy are also examined and modified.

A checklist for the physiological aspects of ergonomics would include the following:

- Machine vibration
- Illumination
- Temperature
- Noise
- Fumes
- Gasses
- Dust

For the handling of material, the following factors would be considered:

- Reach distance
- Back support
- Mechanical aids
- Minimum carrying distance
- Use of mechanical loaders, etc.

Psychological Ergonomics

In psychological ergonomics, humans act as the sensor when they read instruments, check gauges, and react on warning systems. Because of this interaction, gauges must be easy to interpret, and should indicate what they represent.

This interpretation of signals is an important part of ergonomics and ensures that the sensor reads instruments correctly, and takes correct action. The information given must be clear and unmistakable, and therefore the positioning, color, and indication on gauges and controls are important to prevent accidents. One normally expects to open a tap by turning it anti-clockwise and to close it by turning it clockwise. Electrical knobs are normally turned clockwise to increase the current, and anti-clockwise to reduce the current.

A checklist for gauges and switches would be as follows:

- Is there any glare present?
- Are the gauges properly displayed?
- Are warning signals distinct and separate for separate messages?
- Do all the similar gauges read the same way?

- Will the operator know when the gauge is malfunctioning?
- Can the dials and gauges be read quickly and accurately?
- Can any changes or differences on a gauge be easily spotted?

Because of the extent of ergonomic considerations, factors such as illumination, temperature, humidity, and general comfort of the workspace should also be considered. Adequate time should be allowed for the completion of tasks and boring work should be changed to reduce worker fatigue.

Example – Inadequate Ergonomic Design

Employees who worked on a packing and shipping line had to take packed cardboard boxes off the conveyor belt and load them onto pallets on the floor. This involved lifting at an awkward angle, swinging the box from the conveyor to the pallet, bending, and loading the box on the pallet. This was done continuously for almost half a work shift every day. This was a new packing line installation and had only recently been commissioned.

After a week, two employees who complained of low back pain were sent for medical examination and were subsequently booked off from work as a result of their injuries. The other workers on the line complained to management that the new set-up required a lot of manual handling and manipulating of heavy boxes, which were awkward to lift and manipulate.

High-risk conditions – The high-risk conditions were the manual handling requirements of the new conveyor loading station, incorrect positioning of the pallets, and the height of the conveyor.

The root cause analysis is as follows:

Why? – Manual handling of awkward loads required by task (Root Cause – Job Factor).

Why? – Conveyor belt mounted too low, no ergonomic consideration (Root Cause – Job Factor).

Why? – Pallets were stacked on the floor requiring bending while lifting (Root Cause – Job Factor).

Why? – Risk of a new installation not assessed for ergonomics (Root Cause – Job Factor).

Why? – Ergonomic considerations of manual lifting and handling not considered (Root Cause – Job Factor).

Why? – No ergonomic pre-design done (Root Cause – Job Factor).

Why? – Mechanical stacking devices not considered (Root Cause – Job Factor).

Why? – Inadequate ergonomic design (Root Cause – Job Factor).

High-risk behavior – Lifting and twisting heavy and awkward loads repeatedly.

The root cause analysis is as follows:

Why? – Awkward lifting positions required by task (Root Cause – Personal Factor).

Why? – Repetitive lifting motions required by task (Root Cause – Job Factor).

Why? – Lifting and placing heavy boxes repeatedly required by task (Root Cause – Job Factor).

Why? – Bending and twisting motions are required to load boxes (Root Cause – Personal Factor).

Why? – No ergonomic considerations applied to workstation (Root Cause – Job Factor).

WEAR AND TEAR

Factors to consider for the root cause of wear and tear are the deteriorated condition of tools, equipment, facilities, or equipment used beyond normal service life. The abnormal use of equipment, or the overloading thereof, or using the equipment at excessive speed are examples of wear and tear. The equipment should not be used in a different way from what was anticipated when it was purchased or built. Nor should the equipment or facility be used beyond its normal service life. Equipment and tools must be monitored for wear and tear. Equipment used by untrained employees and operated at excessive speed, temperature, or pressure, or operated in any other abnormal way, is deemed to be wear and tear.

If tools and equipment are used for the incorrect purpose, this will contribute to excessive wear and tear and in some cases failure. Compensation should be made for normal wear and tear, and where possible, monitoring of wear and tear should take place.

EXAMPLE – WEAR AND TEAR

While removing a shaft on a cement kiln rotation cylinder, the pipe wrench slipped off the shaft and knocked the employee's knee, causing a minor injury. Upon inspection, the grip on the wrench's jaw's surfaces was found to be almost smooth and offered almost no grip on the shaft at all. The wrench has slipped off the shaft once the worker applied pressure to the handle of the wrench.

High-risk condition – The gripper threads on the pipe wrench jaws were worn. The root cause analysis is as follows:

Why? – Over time the gripper threads had been worn smooth due to the work done with them (Root Cause – Job Factor).

Why? – The wrench had been in service for a long time (Root Cause – Job Factor).

Why? – The threads had worn down with use (Root Cause – Job Factor).

Why? – There is no way to rebuild or repair these gripper threads (Root Cause – Job Factor).

High-risk behavior – The employee was using a worn pipe wrench to remove the shaft.

The root cause analysis is as follows:

Why? –The same tool had always been used for this job in the past (Root Cause – Personal Factor).

Why? – The employee was unaware of the threads being worn smooth (Root Cause – Personal Factor).

Why? – Tools were not examined before each use (Root Cause – Personal Factor).

Why? – The company did not provide all the tools needed and this was the employee's personal wrench (Root Cause – Job Factor).

REGULAR INSPECTIONS

Ideally, tools should be replaced when they show signs of excessive wear. Tools that have been overstressed or otherwise damaged should be replaced promptly. Employees' personal tools should also receive a periodic inspection for wear and tear.

ABUSE OR MISUSE

Based on past work methods were the wrong tools, equipment, or facilities used for convenience, intentionally, or due to personal preference? Were equipment or facilities in a deteriorated or damaged condition due to abuse, misuse, or vandalism?

EXAMPLE – ABUSE OR MISUSE

A farmer sent his farm tractor to the local agents to have the hydraulic pump repaired. The town was some distance from the farm, and it was decided to drive the tractor into town, to the agents. On the way back to the farm, the tractor which was going down a slight decline in the road started to bounce up and down. The driver eventually lost control and the tractor ran off the road. Fortunately, the driver was not injured. The investigation into this accident showed that the tractor had no suspension, and relied on the tires to absorb bumps. On a paved road, the tires could not absorb the shocks and bumps of the hard surface, and this caused the tractor to bounce up and down and the driver lost control.

High-risk condition – The high-risk condition was the tractor driving on a hard paved surface for a long distance.

The root cause analysis is as follows:

Why? – The tractor had no suspension like an automobile (Root Cause – Job Factor).

Why? – The tractor could not absorb the shocks of the road, and this caused the tires to bounce (Root Cause – Job Factor).

Why? – The tractor was not designed for long trips on paved roads (Root Cause – Job Factor).

Why? – The tractor was designed for soil-like conditions and not asphalt roads (Root Cause – Job Factor).

Why? – Misuse of equipment (Root Cause – Job Factor).

High-risk behavior – The high-risk behavior was the decision to send the tractor on a long trip on a paved road.

The root cause analysis is as follows:

Why? – The hydraulic pump needed repair (Root Cause – Job Factor).

Why? – The tractor agents were in town (Root Cause – Job Factor).

Why? – It was convenient to send the tractor in by road (Root Cause – Personal Factor).

Why? – A saving of time and money (Root Cause – Personal Factor).

Why? – Inadequate leadership, decision not based on risk (Root Cause – Personal Factor).

CONCLUSION

Deriving the root causes of an accident involves firstly identifying the immediate accident causes in the form of high-risk behaviors and high-risk workplace conditions. A root cause analysis is then done by using the "Why?" method which involves asking the question "Why?" at least five times for each immediate cause. The answers are the accident root causes.

Sometimes, when questioning, an immediate cause such as a high-risk behavior, the answer uncovers another high-risk behavior or high-risk condition. These immediate causes should again be questioned, over and over, until a deep-seated root cause is uncovered.

A root cause analysis conducted by asking "Why?" is a simple but effective method of deriving the root causes of the high-risk behavior and high-risk workplace conditions. Sometimes, there is an overlap between behaviors and conditions, and job and personal factors, but these should not detract from the fact that the analysis has uncovered a multitude of underlying root causes for the event.

21 Safety Management System (SMS) Controls

INTRODUCTION

The accident sequence is triggered off by a failure to identify the hazards and assess the risks. If the risks arising out of a business have not been identified and assessed, they cannot be managed or controlled. This creates a lack of management control, or poor management control, in the form of weaknesses in the safety management system (SMS), which then triggers off the rest of the accident sequence.

The lack of or weakness in control is depicted by the second domino in the chain of events leading to accidental losses (Figure 21.1).

MANAGEMENT

A manager is anyone who uses management skills or holds the organizational title of "manager." A manager is in control of resources and gets things done through other people by performing the act(s) of management.

SAFETY AUTHORITY, RESPONSIBILITY, AND ACCOUNTABILITY

Occupational health and safety in the workplace starts with, and is led by top management, and is supported by all other levels of management. Therefore, authority, responsibility, and accountability should be clearly defined for all levels of

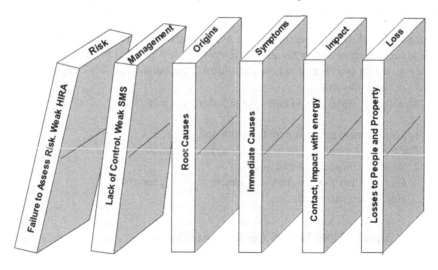

FIGURE 21.1 Lack of control due to a weak SMS.

DOI: 10.1201/9781003220091-24

management, employees as well as for contractors, union officials, and other relevant parties within the organization.

SAFETY AUTHORITY

Authority is the right or power assigned to a manager in order to achieve organizational objectives. Safety authority is formal, specified authority, which gives a manager the authority over health and safety within the organization. All Health and Safety Management System standards, policies, and procedures should clearly define authority, responsibility, and accountability for all levels.

SAFETY RESPONSIBILITY

Safety responsibility is the obligation to carry forward an assigned task to a successful conclusion without accidental loss. Safety responsibility is also an obligation entrusted to the possessor, for the proper custody, care, and safekeeping of property, or supervision of individuals. With responsibility, goes authority to direct and take the necessary action to ensure success.

SAFETY ACCOUNTABILITY

Safety accountability is when a manager is under obligation to ensure that safety responsibility and authority are used to achieve health, safety, and legal standards, without accidental loss. It means being liable to be called on to render an account, or be answerable for decisions made, actions taken, outcomes, and conduct, including workplace accidents. This is the reason that managers take the lead in all accident investigations, and not the safety department.

MANAGEMENT LEADERSHIP

The health and safety of employees at a workplace is the ultimate responsibility of the management of the organization. Even though it is generally accepted that all share a role in health and safety, the ultimate accountability lies heavier with all levels of the leadership.

An effective SMS should contain the following elements:

1. Management commitment to health and safety.
2. Employee involvement in health and safety issues.
3. Worksite inspections and audits.
4. Hazard identification, risk assessment, and risk control.
5. Health and safety training.

FOUR BASIC FUNCTIONS OF MANAGEMENT

It has generally been accepted that a manager's main functions are *planning, organizing, leading* or *directing*, and *controlling*.

All these functions entail the management of employees, materials, machinery, and processes, and should include the management of health and safety across the sphere of all of these main functions. This would include managing the health and safety of all resources, namely, people, material, equipment, and the work environment under the manager's control.

An accident (or high-potential near-miss incident) is an indication that some aspect of management control is missing, has a weakness, or has failed. Accident investigation is a method to identify the failure and recommend measures to improve the controls.

SAFETY PLANNING

Safety planning is to predetermine the probability and possible consequences of accidents and to decide on actions to be taken to prevent these adverse events occurring. One function of safety planning is the recording, analyzing, investigation, and remedying of accident causes.

Safety planning involves:

- Safety forecasting – Hazard identification and risk assessment.
- Setting safety objectives – Setting objectives for the elimination of hazards, and setting standards and parameters for effective accident investigations.
- Setting safety policies – Introducing elements (policies, standards, and procedures) of the SMS, including accident reporting and investigation.
- Safety programming – Prioritizing safety actions, time-lining accident investigations.
- Safety scheduling – Establishing time frames for the health and safety activities.
- Safety budgeting – Allocating funds for mechanical or structural repairs or modifications to eliminate hazards reported through the SMS and accident investigation system.
- Establishing safety procedures – Establishing the accident reporting and investigation procedure, including the training of investigators.

As part of safety planning, the necessary investigation forms, or electronic recording methods, as well as the steps needed to investigate an accident must be established and implemented. These specific investigation systems and methods must be developed, and designated investigators must be trained and appointed.

SAFETY ORGANIZING

Safety organizing is the function a manager carries out to arrange safety work to be done most effectively by the right people. This would also involve allocating employees to investigate accidents, and to do the follow-up on remedial actions taken.

The functions of safety organizing are as follows:

- Integrating health and safety into the organization
- Delegation for safety
- Creating safety relationships

- Establishing safety authority
- Creating safety responsibility
- Creating safety accountability

INTEGRATING HEALTH AND SAFETY INTO THE ORGANIZATION

Integrating health and safety into the organization is the work a manager carries out to allocate safety work to all levels of employees. The more the SMS is integrated into the organization, the better the safety culture becomes. Only once the safety activities of an organization are blended into the day-to-day operations of the organization, and not perceived as separate functions, can safety culture change takes place. Once complete integration is achieved, a positive, proactive safety culture will exist.

DELEGATION FOR SAFETY

Safety delegation is what a manager does to entrust safety responsibility and give safety authority to his subordinates while at the same time creating accountability for safety achievements.

CREATING SAFETY RELATIONSHIPS

Creating safety relationships is work done by a manager to ensure that safety work is carried out by the team with utmost co-operation and interaction among team members. The ongoing SMS requires participation, support, and action from all levels within the organization and cannot be left as one person's, one department's, or one manager's responsibility.

ESTABLISHING SAFETY AUTHORITY

Safety authority is the total influence, rights, and ability of the position, to command and demand safety actions. Management has ultimate safety authority, therefore, is the only echelon that can effectively initiate, implement, and maintain an effective SMS. Leadership has the authority to implement and maintain health and safety and also the authority to take necessary remedial measures if there are deviations from accepted safety practices and norms.

CREATING SAFETY RESPONSIBILITY

Safety responsibility is the safety function allocated to a position in relation to the authority of the post and it determines the level of safety accountability of that post. It is the duty and function demanded by the position within the organization. This lies with all levels of management as well as with employees. The higher the management position, the higher the degree of safety responsibility.

One cannot be held accountable for something over which one has no authority. The degree of safety accountability is also apportioned to the degree of safety authority. Job descriptions are vital management tools and should clearly define the safety authority, responsibility, and accountability for all jobs and all levels within the organization.

CREATING SAFETY ACCOUNTABILITY

Creating safety accountability is when a manager is under obligation to ensure that safety responsibility and authority is used to achieve both safety and legal safety standards. Employees too have safety accountabilities but in proportion to their safety authority.

Leadership is accountable to manage a SMS and to provide the necessary infrastructure, environment, and training to enable the system to work. Employees should be held accountable for participating in the system and for following the prescribed standards of the system.

Management at all levels is then held accountable to rectify the problems identified by the safety system, including accident investigations, and to ensure that the high-risk acts or conditions highlighted by the system are rectified before a future loss event occurs.

SAFETY LEADING

Safety leading is when managers take the lead in health and safety matters, make safety decisions, and always set the safety example. The functions of safety leading are as follows:

- Making safety decisions
- Safety communicating
- Motivating for safety
- Appointing employees
- Developing employees

MAKING SAFETY DECISIONS

Making safety decisions is when a manager makes a decision based on safety facts presented to him or her. During an accident investigation, there will be a multiplicity of decisions that need to be taken.

SAFETY COMMUNICATING

Safety communicating is what a manager does to give and get understanding on safety matters. This is one of the key components in the SMS. This function would entail the creating of various safety communication channels from the workplace to management and from management to the workplace. Regular feedback to the workforce on safety matters such as the results of an accident investigation helps foster safety communication and reporting by employees and informs management about conditions on the shop floor.

MOTIVATING FOR SAFETY

Motivating for safety is the function a manager performs to lead, encourage, and enthuse employees to take action to improve safety. Safety incentives should never be in the form of cash and should never be linked in any way to injuries. They should rather be small tokens of appreciation given to employees who take positive safety actions. Sometimes, even a simple handshake is appreciated.

Examples of safety actions that can be acknowledged:

- Maintaining the housekeeping standards.
- Frequent reporting of near-miss incidents.
- Safety suggestions submitted.
- Wearing of personal protective equipment.
- Attendance at safety committee meetings, etc.

Acknowledging employees is vital to the success of the system. The power of this small gesture of thanks and recognition cannot be overstated in its importance in promoting safety at the workplace.

APPOINTING EMPLOYEES

Appointing employees is a management function where management ensures that the person is both mentally and physically capable of safely carrying out the work for that position. Hiring the correct person for the task is a prime consideration for the prevention of accidents. The success of coordinating the SMS will indirectly depend on the support it receives from the safety department and this depends on the quality of health and safety staff initially appointed.

DEVELOPING EMPLOYEES

A manager develops employees by helping them improve their health and safety knowledge, skills, and attitudes. Management has to ensure the health and safety staff are up to date with the latest trends in safety and risk management and that there is an ongoing self-development program in place. Training employees in the techniques of accident investigation is developing employees.

SAFETY CONTROLLING

Safety controlling is the management function of identifying what must be done for health and safety (implementing and maintaining the SMS), inspecting to verify completion of work (plant inspections), evaluating, (SMS audits), and follow-up with safety action. A weakness in safety management control is what triggers the accident immediate causes. Controlling is the most important safety management function!

The control function has seven steps:

1. Hazard identification, risk assessment, and safety actions to be taken.
2. Set standards of performance measurement (*what* must be done).
3. Set standards of accountability (*who* must do *what*, by *when*).

4. Measure against the standards.
5. Evaluation of conformance.
6. Corrective action.
7. Commendation.

1. Hazard Identification, Risk Assessment, and Safety Actions to be taken

Based on hazard identification and risk assessments, a manager lists and schedules the work needed to be done to create a healthy and safe work environment, and to eliminate high-risk behaviors. This would mean the introduction of a suitable structured SMS, based on the world's best practice.

A health and safety management system is defined as *Ongoing activities and efforts directed to control accidental losses, by monitoring critical health and safety management system elements on an ongoing basis.*

2. Set Standards of Performance Measurement (*What* must be done)

Safety standards are "measurable management performances." Standards are set for the level of work to be done to maintain a healthy and safe environment, free from actual and potential accidental loss. Standards are established in writing for all the health and safety management system elements. Without standards, the management system has no direction, nor are safety expectations established. (If you don't know where you're going, any road will take you there.)

3. Set Standards of Accountability (*Who* must do *what*, by *when*)

Management now sets standards of accountability by delegating authority to certain positions for ongoing safety work to be done. Coordination and management of the SMS need to be allocated to certain departments and individuals, and these standards dictate *who* must do *what,* by *when,* to implement and manage the system.

4. Measure against the Standards

By carrying out safety inspections, the actual condition of the workplace and the ongoing activities of employees are now measured against the accepted safety standards. Physical inspection of the workplace will confirm if risk control measures recommended by accident investigations have been implemented.

If there is no formal system of measurement, then management does not know how well the SMS system is doing, compared to its own standards and best practice.

5. Evaluation of Conformance

Depending on which measurement method is used, the conformance results are now quantified in the form of a percentage allocated, marks given, or a ranking established. Safety audits, both internal and external, evaluate compliance with an organization's SMS standards, and scores then indicate whether there is a deviation from the prescribed standards. Missing, weak, or non-functioning control elements of an SMS invariably result in an accident.

6. Corrective Action

The amount of corrective action will be proportional to the degree of deviation from standards. Corrective action may involve enforcing the safety standards and taking the necessary action to regulate and improve the methods.

Corrective actions are the defining stages in an accident investigation system, and are where the rubber meets the road. The investigation will indicate what immediate and root causes triggered the event. The only way to prevent a recurrence of a similar event, which may result in an accident, is to take corrective preventative actions. This may mean fixing the high-risk condition or correcting high-risk behavior, or a combination of both, and then rectifying their root causes.

Standards are established for these corrective actions and they must state *who* must do *what*, by *when*, in order to get the situation rectified. Corrective actions must be positive, time-related, and be assigned to responsible people.

7. Commendation

Commendation is when a manager pays compliment and expresses gratitude for adherence to achievement of pre-set health and safety standards. If employees are not recognized for participating in the company SMS, their enthusiasm will soon wane. Recognizing and reporting hazards must become a part of the company's culture, and management's involvement, support, and participation in the system are vital.

FIGURE 21.2 The four basic safety functions of management. [From McKinnon, Ron C. 2012. *Safety Management, Near Miss Identification, Recognition and Investigation.* Boca Raton: Taylor and Francis. With permission.]

CONCLUSION

An accident is the final result of a failure to identify hazards, assess the risks, and implement risk mediation methods in the form of health and safety controls within the SMS. The management risk assessment and control functions are, therefore, vital to prevent accidents. Once they are implemented, the chain of events resulting in an accident is less likely to be triggered.

Section IV

Remedial Measures to Prevent a Recurrence

22 Failure to Identify Hazards and Manage Risks

INTRODUCTION

Invariably, most occupational accidents can be traced back to a failure to identify the hazards and assess the potential of these to cause a loss. After an accident occurs, applicable control mechanisms are usually implemented, but these should have been in place before the accident occurred.

This failure of the hazard identification and risk assessment (HIRA) system leads to a breakdown in management controls, in the form of weaknesses in the safety management system (SMS). This creates root causes (personal and job factors) which, in turn, lead to immediate accident causes (high-risk behavior and high-risk workplace conditions), the subsequent energy exchange (exposure, impact, or energy exchange), and the consequent losses.

HAZARD IDENTIFICATION AND RISK ASSESSMENT

HIRA is one of the most important elements (programs) within the health and safety management system. If applied diligently, HIRA could reduce the number, frequency, and severity of adverse events occurring in a workplace.

THE PURPOSE OF HIRA

The purpose of HIRA is to identify hazards, both high-risk behaviors and conditions, and evaluate the probability and severity of injury, illness, or damage arising from exposure to these hazards. The goal of HIRA is the elimination of the risk or implementation of control measures to mitigate the risk. This should be an ongoing process and will contribute greatly to the reduction of accidents. Identifying hazards, assessing their risk, and eliminating them will prevent the domino effect of an accident from being triggered.

HIRA for the work at hand should be an ongoing process as laid down in the standards of the SMS. The risk assessments should include identifying which company policies, standards, procedures, and processes apply to the workplace situations.

SOURCES OF HAZARDS AND HAZARD BURDEN

Sources of hazards should be identified and documented in a risk register that is modified and updated regularly. This will direct the design and requirements of the SMS, as the system must be focused on reducing risks arising from the workplace and its activities.

DOI: 10.1201/9781003220091-26

When measuring the hazard burden, the following questions should be asked:

- What are the hazards associated with the business activities?
- What is the significance of the hazards (high/low)?
- How does the nature and significance of the hazards vary across the different areas of the organization?
- How does the nature and significance of the hazards vary with time?
- Is the organization succeeding in eliminating or reducing hazards?
- What impact do changes in the business have on the nature and significance of hazards?

HIRA OBJECTIVE

The objectives of the HIRA process are as follows:

- To identify all the pure risks within the organization which are connected to the operation (*hazard identification*).
- To do a thorough analysis of the risks taking into consideration the frequency, probability, and severity of consequences (*risk analysis*).
- To implement the best techniques for risk reduction (*risk evaluation*).
- To manage the risk (*risk control*).
- To monitor and re-evaluate the risk on an ongoing basis (*re-assessment*).

TYPES OF WORKPLACE HAZARDS

SAFETY HAZARDS

These are hazards that create high-risk workplace conditions. Examples include exposed electrical wires that could shock or electrocute, or a slippery walkway that might result in someone slipping and falling.

BIOLOGICAL HAZARDS

Biological hazards include viruses, bacteria, insects, animals, and other sources that can cause adverse health impacts. These include blood and other bodily fluids, mold, vermin, sewage, dust, and harmful plants.

CHEMICAL HAZARDS

Chemical hazards are hazardous substances that can cause harm to employees and the work environment. These hazards can result in both health and physical impacts, such as respiratory system irritation, blindness, skin irritation, corrosion, as well as explosions and fire.

PHYSICAL HAZARDS

Physical hazards are environmental factors that can harm employees without necessarily touching them, including noise, vibration, radiation, and pressure.

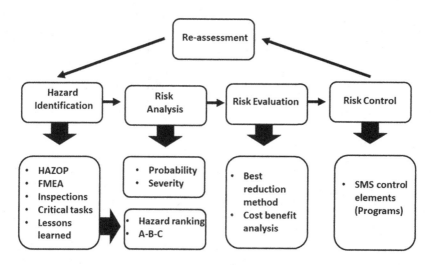

FIGURE 22.1   The HIRA process.

ERGONOMIC HAZARDS

Ergonomic hazards are a result of physical factors that can result in musculoskeletal injuries. Manual lifting and handling, a poor workstation setup in an office, poor posture, and repetitive motion tasks are examples.

PSYCHOSOCIAL HAZARDS

Psychosocial hazards include those that can have an adverse effect on an employee's mental health or well-being. These include workplace stress, sexual harassment, victimization, and workplace violence.

HIRA PROCESS

Root causes are the reasons why high-risk acts are committed, and why high-risk conditions exist in a workplace. Ideally, these need to be identified before the occurrence of an adverse event and rectified by eliminating their causes. Once the hazards are identified, their root causes can be eliminated. The HIRA process includes the following:

- Identifying tasks done before, during, and after the work.
- Identifying potential hazards for each task.
- Determining the level of risk, using probability and consequence on the risk matrix.
- Identifying controls to be put in place, with consideration given to how effective and adequate the proposed controls would be.

IDENTIFYING TASKS

In any organization, there are a number of tasks that are carried out, and a number of processes and procedures that take place continuously. Before the hazards of these activities can be identified, these tasks and processes need to be identified. Sources of energy used must be identified, as well as the inputs and outputs of the business. Production line functions should be identified, and tasks carried out by workers listed. This in itself is a mammoth task but needs to be done so that all movements and activities at the workplace are identified. Only once the tasks and processes are identified and noted, can the hazard classification begin.

IDENTIFYING POTENTIAL HAZARDS

The first step of a risk assessment is the identification of all possible hazards. A hazard is a situation that has the potential for injury or damage to property or harm to the environment. It is a situation or action that has the potential for loss.

Via discussions with employees involved in the listed processes, a hazard identification and ranking exercise can be conducted. Although this may be a preliminary assessment, it will nevertheless give a ranking of the hazards for further assessment. Meetings, discussions, and brainstorming sessions should be held which will help to identify the potential hazards within the processes and tasks.

There are numerous hazard identification methods and techniques. The two main techniques are the *comparative* (non-scenario based) and the *fundamental* methods (scenario-based).

Comparative techniques include the following:

- Checklist analysis
- Safety reviews
- Preliminary hazard analysis
- Past experience, etc.

Fundamental techniques include the following:

- "What if" analysis
- Hazard and Operability studies (HAZOP)
- Failure Mode and Effect Analysis (FMEA)
- Failure Mode, Effect, and Critical Analysis (FMECA)
- Fault Tree Analysis
- Event Tree Analysis, etc.

Once the hazards are identified, they can be roughly grouped according to their hazard classification. A simple method of hazard ranking is the A, B, C method:

A-Class hazard – Has the potential to cause death, major injury, or extensive business interruption.
B-Class hazard – Has the potential to cause serious, non-permanent injury, and minor disruption of the business.
C-Class hazard – Has the potential to cause minor injury and non-disruptive business interruption.

Once hazards have been prioritized in a hazard-ranking exercise, which follows the hazard identification process, the next step in the HIRA process is the assessment of the risks posed by those hazards.

DETERMINING THE DEGREE OF RISK (RISK ASSESSMENT)

Risks cannot be properly managed until they have been assessed. The process of risk assessment is the evaluation and quantification of the likelihood of undesired events occurring, and the likelihood of injury and damage that could be caused by the risks. It also involves an estimation of the results of adverse events occurring.

One of the biggest benefits of risk assessment is that it will indicate where the greatest gains can be made with the least amount of effort, and which activities should be given priority. Management now has a prioritization system based on sound risk assessment practices.

A simple risk matrix is used to determine the potential for loss and potential severity of that loss, should it occur. The risk matrix will indicate which risks are tolerable (some residual risk will always remain) and what risks need urgent attention (Figure 22.2).

IDENTIFYING THE CONTROLS

Once the three steps of HIRA are completed, risk control is then implemented. Risk control can only be instituted once all hazards have been identified and all risks quantified and evaluated. The hierarchy of control provides a logical prioritization of hazard elimination.

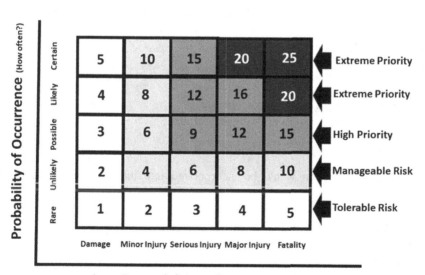

FIGURE 22.2 A risk matrix showing the prioritization of risk mitigation.

Hierarchy of Control

There should be multiple layers of controls protecting employees from hazards and associated risks. The types of controls could include

- Elimination – Which would mean changing the way the work is to be done.
- Engineering controls – Such as the re-design of a worksite, equipment modification, and tool re-design.
- Administrative controls – Such as a change of work methods and re-scheduling of work.
- Health and safety system controls – Such as health and safety policies and standards, work procedures, training, and personal protective equipment (PPE) (SMS elements).

SMS CONTROL ELEMENTS (PROGRAMS)

The SMS contains a number of safety programs, processes, and actions, defined by SMS standards, which are designed to provide ongoing hazard identification, and elimination actions. This is the main purpose of an SMS. These elements of the SMS can only be implemented once the hazards are identified. Examples of the SMS elements (programs) for the different types of hazards are as follows:

SAFETY HAZARDS

The SMS would include the following hazard reduction programs (SMS elements):

- Good housekeeping standards and procedures
- Ladder safety
- Structures, buildings, and floors
- Demarcation of walkways
- Stacking and storage
- Machine guarding
- Electrical safety, etc.

BIOLOGICAL HAZARDS

The biosafety program would be incorporated into the SMS and includes the following:

- Blood-borne pathogen program
- Food safety program
- Vermin control
- Workplace cleaning program
- PPE issue and control
- Personal hygiene program, etc.

CHEMICAL HAZARDS

The SMS would include a complete chemical safety program (SMS element) consisting of the following:

- Purchasing specifications
- Material Safety Data Sheets availability
- Hazard and Communication process (HazCom)
- Respiratory program
- PPE requirements
- Hazardous substance, use, storage, and disposal program
- Emergency equipment provision
- Training, etc.

PHYSICAL HAZARDS

The SMS would include programs (SMS elements) covering the following:

- Hearing conservation program
- Work at height program
- Ventilation control
- Illumination control
- Confined space entry procedures
- Monitoring procedures
- Radiation control, etc.

ERGONOMIC HAZARDS

The SMS would include a complete workplace ergonomic program consisting of the following:

- Management leadership and employee participation
- Hazard identification and information
- Job hazard analysis and control
- Training in ergonomics
- Musculoskeletal Disorders Management
- Ergonomic program evaluation

PSYCHOSOCIAL HAZARDS

The following programs (SMS elements) would address workplace psychosocial hazards:

- Workplace cardinal rules
- Human resources policies
- Company discipline procedures

- Working conditions
- Labor relations
- Counselling programs

ACCIDENT CAUSATION

There are numerous accident causation theories, but one thing is agreed, is that accidents (excluding acts of nature such as flashfloods, earthquakes, and other natural phenomena) are normally a series of small blunders that result in some form of loss or disruption. They are caused and can therefore be prevented. Managements' ongoing process of hazard identification, risk assessment, and risk mitigation via the SMS stabilizes the row of dominoes and prevents the accident domino effect culminating in a loss.

SMS CONTROLS

If an accident occurs, it is because one or more control mechanisms are not in place, have a weakness, or have failed. More than likely, the root cause analysis will indicate where the failure occurred and trace it back to a breakdown in the HIRA process. The object of accident investigation is to determine these root causes and pose solutions, to prevent them arising and causing accidents in the future.

The organization's health and safety management system would consist of a number of control mechanisms (SMS elements or programs), which are aimed at identifying hazards, classifying them, and reducing their risks as far as is reasonably practicable. Work standards, procedures, and policies within the SMS require an ongoing and proactive application to reduce the probability of an accident occurring.

CONCLUSION

An effective accident investigation will recommend one or more or a combination of control remedies to prevent similar accidents occurring in the future. The investigation should also uncover where the ongoing HIRA system failed, and recommend the necessary control measures to ensure it does not fail in the future.

23 Risk Control Remedial Measures to Prevent Accident Recurrences

OBJECTIVE OF HEALTH AND SAFETY IN THE WORKPLACE

The objective of health and safety management is to create a workplace free from hazards and an environment that poses minimal risk to employees in that workplace. This would include protecting equipment and machinery from accidental damage, and the environment from any form of pollution or harm.

MANAGEMENT'S RESPONSIBILITY

It is management's prime responsibility to implement and maintain effective systems, standards, policies, and procedures to ensure that the workplace remains free from hazards. To do this, management should initiate a formal Health and Safety Management System (SMS) with the local legal requirements as the minimum standards, and local and international health and safety standards as the desired objective to be achieved.

EMPLOYEES' RESPONSIBILITY

It is the employees' responsibility to adhere to the health and safety rules of the organization, participate in the safety process, and report hazards noticed in their work area. Employee participation in health and safety can be facilitated via various elements (programs) within the SMS such as

- The safety committee system
- Reporting of hazards
- Reporting of near-miss incidents
- Safety suggestion scheme
- Appointments as health and safety representatives
- Accident investigation participation
- Incident recall system
- Health and safety training
- Plant inspections
- SMS audit participation

DOI: 10.1201/9781003220091-27

ACCIDENTS ARE INDICATORS

Accidents and high-potential near-miss incidents are indicators that one or more of the SMS control measures is missing, inadequate, has a weakness, or has failed. Effective accident investigation will investigate the event and uncover these weaknesses so that the remedial action of risk control can be put in place to implement, improve, or otherwise ensure that the system does not fail in the future. If these measures had been in place initially, the event would not have occurred, as they would have prevented the chain of events occurring.

To be effective, the process of recommending and implementing remedial measures must fix the problem. These measures must focus on eliminating the root causes identified by the investigation.

HIERARCHY OF CONTROL

When implementing the remedial risk controls, the hierarchy of hazard controls should always be considered. They are

- Elimination controls – This would mean physically removing the hazard.
- Engineering controls – Such as the re-design of a worksite, equipment modification, and tool re-design. This is replacing the hazard.
- Administrative controls – Such as a change of work methods and re-scheduling of work.
- Health and safety system controls – Such as health and safety policies and standards, work procedures, training, and personal protective equipment (PPE) (SMS elements).

ACCIDENT REMEDIAL MEASURES (RISK CONTROL)

An accident investigation would propose a number of risk control actions to be taken to prevent the recurrence of the same or similar event occurring. These actions would be based on SMS control elements (programs) that were missing, were weak, or failed.

A list of alternate risk control measures should also be compiled by the investigators. These alternate risk control measures should be evaluated on the basis of their ability to prevent accident recurrences, and whether or not they can be successfully implemented.

Some remedies may be easier to implement than others. Some may be long term, others short term. Cost, and time to implement, should be weighed against the degree of risk reduction obtained. This analysis of the remedies should be done before the final action plan is implemented. When prioritizing the remedial measures, the following priorities should be considered:

- Remedial measures that eliminate the risk entirely, such as using a safer substitute product or process.
- Remedial measures that combat the risk at source, such as the installation of a machine guard.

- Remedial measures that minimize the risk of relying on human behavior. These are administrative controls and may involve safe work procedures, shift rotation, or similar changes.
- The cost of and benefit derived from implementing the controls.

COST-BENEFIT ANALYSIS

Management decisions are mostly based on cost and return on investment. The post-accident proposed risk control measures are also subjected to the same question of cost and benefit. Accident investigation would not be effective if the remedial actions proposed were unreasonable in terms of cost, time, and risk reduction achieved.

The cost-benefit analysis involves a comparison of the cost of risk reduction measures, with the risk factor cost of the accident prevented. It estimates the value of a preventative action by considering the efficacy (how much control will result?) with the feasibility (how acceptable is it?) with the efficiency (the best bang for the buck). The remedies must be cost-effective, practical, and effective, in that, they must solve the problem. In all workplaces, a degree of risk is tolerated. These risks are tolerated once they have been assessed as being *as low as is reasonably practicable.*

ROOT CAUSES ELIMINATED

The recommendations for corrective remedial action arising from the investigation must address the root causes of the accident. These are the causes of the event, and any remedial action in the form of risk control measures must be aimed at treating these root causes.

In the accident scenario where an employee had to climb up a fixed ladder attached to a tank to read the pressure gauge, the root cause of the accident, and the proposed risk control remedies are as follows:

EXAMPLE OF AN ACCIDENT (BUSINESS INTERRUPTION)

An employee had to climb up a vertical fixed ladder attached to a tank to take the pressure readings on a gauge at the top of the tank. If the pressure went over the red line, he was to contact the control room and warn them. During one shift, there was an overpressure warning alarm in the control room, and the plant had to shut down the one boiler to prevent an explosion. No warning had been received from the employee.

Upon investigation, it was found that the employee had a bad knee and had difficulty in climbing the ladder every four hours as required. He subsequently failed to read the gauge, and this led to a situation where he did not warn the control room of a high-pressure reading. The control room's overpressure warning alarm activated, and the one boiler had to be shut down. He was physically incapable of doing the task.

ROOT CAUSE ANALYSIS – HIGH-RISK BEHAVIOR

High-risk behavior – The high-risk behavior was not reading the gauge and not warning the control room.

The root cause analysis is as follows:

Why? – The employee did not read the gauge every four hours as his knees were sore and he could not climb the ladder (Root cause).

Proposed Risk Control Remedy – The elimination of having to read the gauge by having a pressure sensor wired directly to the control panel (Ideal solution).

Why? – The employee had weak knees due to sport injuries in the past and had extreme difficulty climbing the ladder (Root cause).

Proposed Risk Control Remedy – Implementation of strict pre-employment examinations in conjunction with employee job specifications for all types of positions within the company. High-risk operations to be prioritized. This is to ensure that the correct employees are selected for the correct positions.

Why? – No warning was sent to the control room (Root cause).

Proposed Risk Control Remedy – Moving the gauge to an easily readable position. Pre-employment selection to ensure a physically capable employee selected for the job.

Root Cause Analysis – High-risk Conditions

High-risk conditions – The high-risk workplace conditions were the incorrect positioning of the gauge and relying on a worker to issue a warning to the control room.

The root cause analysis is as follows:

Why? – The gauge was incorrectly placed (Root cause).

Proposed Risk Control Remedy – New plant and equipment to undergo scrutiny using hazard identification methods such as HAZOP (Hazard and Operability Study), FMEA (Failure Mode and Effect Analysis), and FMCA (Failure Mode and Criticality Analysis). Recommended implementation of a change management system. The gauge would have been identified as a critical part and its positioning (and possible direct linking with the control room) would have been questioned in the design stage.

Why? – The position of the gauge was never specified in the design specifications (Root cause).

Proposed Risk Control Remedy – Implementation of a change management system. Safety and ergonomic considerations to be included in the specifications of new plant and machinery.

Why? – Purchasing specifications did not include ergonomic factors (Root cause).

Proposed Risk Control Remedy – Ergonomic considerations by a qualified ergonomist to be included in all new designs.

Why? – The company did not have a standard on ergonomics in the workplace (Root cause).

Proposed Risk Control Remedy – An ergonomic standard to be written and incorporated into the SMS. This standard includes ergonomic specifications for plant, machinery, and workstations, training and ergonomic surveys, studies, and inspections.

Why? – The employee was not required to undergo a hiring vocational adaptability test upon hiring (physical examination) (Root cause).

Proposed Risk Control Remedy – Job specifications to be introduced with pre-employment medical examinations.

Why? – There was no company standard or requirement for worker job specifications (Root cause).

Proposed Risk Control Remedy – A job specification and pre-employment examination protocol for each work position to be introduced. This will indicate the physical and mental requirements of the positions and are to be used with pre-employment medical examinations to ensure the correct person is placed in the correct job.

Why? – This was not deemed necessary by the company (Root cause).

Proposed Risk Control Remedy – The company should review all work positions and compile a list of physical and mental requirements for these positions. Medical testing and suitability examinations should then be prescribed for each position and issued to appointed medical examiners.

PROPOSED RISK CONTROL REMEDY CONCLUSION

As a result of the investigation, it was found that a hazard identification review at the planning stage would have identified the critical gauge was mounted in the wrong position. The possibility of connecting this gauge directly to the control room was not considered. A *management of change* process at the design stage would have also identified this. This review would include ergonomic considerations related to new plant and equipment. Further, it was found that a critical procedure relied on an operator. The company had no employee job specifications and therefore an employee who was (occasionally) physically incapable of reading the gauge had been appointed.

The ideal remedy would be to mechanically link the gauge readings directly with the control room, eliminating the reliance on an operator. Should this not be possible, or financially feasible, the correct operator should have been selected by using an employee job specification to guide the medical examiner as to what the physical and mental requirements of the job were so that a suitable employee could have been selected. The SMS programs (elements) that were not incorporated were the *safe management of change* process, the *employee job specification* requirements, and the *ergonomic program*. If these programs were incorporated into the existing SMS, the accident would not have happened. A description of the three SMS programs (elements) that were missing is as follows.

SAFE MANAGEMENT OF CHANGE (CHANGE MANAGEMENT)

Change to processes, procedures, and layout at the workplace often brings progress. However, ineffective management of change has been the cause of many major accidents in the past and creates the potential for future accidental events. Unmanaged changes increase risks, that if not properly managed, may create conditions that could lead to injuries, property damage, or other losses. The main types of change within an organization are as follows:

- Changes to infrastructure
- Changes to processes, equipment, or products
- Changes in personnel
- Use of different materials
- Changes brought about by the health and safety management system

Ideally, no modification should be made to any plant, equipment, control systems, process conditions, operating methods, or health and safety procedures, without authorization from a responsible manager. In some cases, third-party approvals may be necessary.

The hazards encountered and risks posed by these changes should be assessed. The necessary controls should then be put in place before the change, to reduce the probability of the change resulting in accidental loss.

The Safe Management of Change Process (Change Management)

There are a number of ways to implement the safe management of change. All achieve the same objective of ensuring that the change does not create additional risks, is of benefit to the organization, its employees and processes, and is communicated to all. A change management checklist and flowchart are vital in change management. An example of the change management process is as follows:

- Identify the change – The change must be identified and clearly defined. Departments, machinery, or processes that will be affected must be listed and consulted.
- Risks and rewards of the change – The hazards created by the change must be identified and their risks assessed. The benefits of the change should be clear and measurable. Affected departments should use this data to approve or reject the change.
- Change approval – Most change initiatives need approval from the applicable authority level. Budgets need to be established, as well as a plan of action which is approved by all affected parties. Once approved, a change management document should be opened.
- Change team – In some instances, a change management team or committee may be needed to manage the change and give feedback on the progress of the change.
- Communicating the change – Before any change takes place, the change should be communicated to all interested and affected parties. The benefits of the change should be stated as well as the commencement and completion date of the change.
- Risk mitigation – Any risks identified as a result of the change should be identified before the change takes place, and risk control measures instituted to mitigate these risks. In all change interventions, the benefits must outweigh the risk.
- Implement the change – The change should be implemented by following a plan of action, which allocates authorities, responsibilities, and accountabilities to bring about the change. The change management form will guide the change owners (project managers) through the correct procedure. Timeline should be established as well as budget requirements.
- Training – Depending on the change, additional training of employees or contractors may be needed to enlighten them on the changes. All affected parties should undergo this update training.

- Confirm effectiveness of change – Once the change is completed, data should be collected that quantifies the efficiency of the change.
- Follow-up – A follow-up on the change is the final step of the change management process. This would check that the change was implemented according to plan, and that the change has created the benefit envisioned in the planning stage. This step would include the final sign-off of the change document.

AN EMPLOYEE JOB SPECIFICATION PROGRAM (SMS ELEMENT)

To enable an organization to function safely, an all-out endeavor should be made to get the correct employee to do the work for which he or she is best suited. This would ensure placing the correct person in the correct job so that there is no mismatch. The possibilities of accidents can be reduced by selecting employees with the correct attributes for the work that they are to perform.

EMPLOYEE JOB SPECIFICATION

An employee job specification is a detailed description of both the physical and cognitive attributes required of a person to fill a specific work function.

The specification specifies the physical and cognitive aptitudes and attitudes required for the person who is to do the work. The job specification can be referred to as a checklist to ensure that the correct person is selected for the correct job.

JOB SPECIFICATION CHECKLIST

All of the variables, employee specifications and requirements, should be compiled into a checklist and used when selecting employees. The person requisitioning the employee should complete the checklist. This checklist will then form an outline of the type of person required for a specific job.

The employee specification should be used for the correct selection and training of employees, and should be taken into consideration when employees are transferred or promoted. Obtaining the correct person in the correct job will reduce the chances of frustration, mismatching, and stress, which could lead to accidents.

AN EXAMPLE OF AN EMPLOYEE JOB SPECIFICATION PROGRAM STANDARD (SMS ELEMENT)

Pre-employment Medical Examination

Before employing a person, he or she should attend a pre-employment medical examination and be interviewed by a selection panel. The objective of the medical examination will be to ensure that the person has the physical attributes and requirements specified for the type of work that he or she has applied for.

The medical examination will examine the applicant to ensure that the person possesses the necessary physical requirements for the post, and that he or she is in good health and is capable of fitting into the work environment.

Physical, Health, and Cognitive Requirements

The employee specification would list the various physical, health, and cognitive requirements demanded by the particular job. The person requesting that the position be filled would indicate to the medical department what specific attributes are required.

These could include eyesight, and the medical practitioner may be required to check depth perception, the vertical and lateral phoria, color vision, and also hearing acuity. This is an important attribute, especially for vehicle and equipment operators.

To ensure that the candidate can heed warning signs and signals, his or her hearing acuity should be tested and a general overview of the person's health examined. Many organizations require drug testing for all new employees.

If the position calls for physical manual handling, the specification may require a person to be of a suitable build, a certain height, and a certain weight.

Personal Protective Equipment

The employee specification will list the items of PPE that the employee might have to wear. The medical examination will determine whether or not the person is able to wear safety shoes, hard hat, safety glasses, or whatever else may be required during the work. If respiratory protection is to be worn, the employee would require a medical certificate stating that they can wear a respirator. Respirator test fitting would then follow.

Educational Qualification and Experience

Depending on the position being filled, the person requesting the manpower will specify what general education is required, what degrees or diplomas are required, and how many years of work experience are required. The general intelligence of the person may be tested as well as the applicant's comprehension and memory. If needed, the specification will list other specific requirements for the position.

Special Aptitudes

When completing the employee specification, special aptitudes required for the job may be listed. These aptitudes could include mechanical ability, manual dexterity, including hand and eye coordination, or hand, eye, and foot coordination. Reaction ability may be required and the ability to speak local languages may also be specified.

Personality

If a person is to be placed in charge of other workers, they must fulfill certain personality requirements. The job specification may list certain leadership requirements such as communication skills, flexibility, initiative, tenacity, and ability to train others. A supervisor, or leader lacking in some of these attributes, may create a situation where friction and stress occur in the workplace, which in turn could give rise to accidents.

Physical Work Conditions

The employee specification will also specify special circumstances under which the person may work, such as the working hours, the physical work conditions such as heat, cold, noise, and special activities (climbing, standing, and bending). The incumbent may be required to shift work, overtime, or to be on standby over weekends.

Physical work conditions could include working in a noise zone, dirty, dusty areas, a hot or cold environment, or may involve working with heavy equipment or other hazardous substances such as explosives. The task may even be repetitive and monotonous, which would require a certain disposition.

WORKPLACE ERGONOMIC PROGRAM (SMS ELEMENT)

Ergonomic-related injuries include injuries related to lifting, pushing, pulling and holding, carrying, or throwing objects. An important step to prevent these types of injuries is to implement an ergonomic safety program, which is the science of adjusting the job to fit the body's needs. Ergonomics provides injury prevention solutions that are simple and relatively inexpensive. Some of these solutions include requiring frequent short breaks for workers assigned repetitive motion tasks, providing manual or mechanical lifting equipment, and varying worker's tasks.

ERGONOMICS DEFINED

Ergonomics is the general term used to identify the field of activity aimed at the matching of machines, equipment, and the environment to humans in such a way that optimal performance can be achieved. The main goal of ergonomics is the reduction of errors and injuries caused by repetitive or awkward motions, and the incorrect matching of the worker to the job.

By applying ergonomics, the comfort of the workplace is improved, and the reliability of humans as a sensor in the worker-machine cycle is also improved. Improving the ergonomics helps reduce the need for training as the relationship between the person and his or her work environment is also improved. Applying ergonomics reduces errors and improves safety.

ERGONOMIC INSPECTIONS AND RISK ASSESSMENTS

A specifically planned inspection should be carried out using an ergonomic checklist. This inspection will be aimed at examining the availability and condition of seating at desks, tables, and worktops, as well as investigating the accessibility of tools, gauges, sample points, foot holes, platforms, etc.

The ergonomic checklist should include all work situations and the method of handling materials, and will also encompass inspecting of gauges, signals, and other warning devices. Once completed, the ergonomic risks identified should be assessed and prioritized for action.

AN ERGONOMIC SAFETY PROGRAM

An ergonomic safety program would cover three main areas:

- Anatomy
- Physiology
- Psychology

Anatomy is concerned with the dimensions of the human body and its variations, and also includes biomechanics, or the forces that can be applied by the body under certain conditions.

Physiology includes both work physiology and environmental physiology. Environmental physiology is the effects of the physical environment on the workplace, and work physiology concerns the expenditure of energy.

Psychology includes skill psychology and occupational psychology. Skill psychology is concerned with the mental activity of information receiving, processing, and decision taking. Occupational psychology concerns individual differences, efforts required, and training. Humans interact with machines by reading gauges and controls. They process this information and give the machine further instructions. Psychological aspects of ergonomics include worker satisfaction and the comfort that the person feels in their job.

Anatomical Ergonomics

The first area of ergonomics, anatomy, concerns the physical facilities such as seating, height of controls, body posture, muscular strength, and body movement. People have different body sizes and shapes, and ergonomics endeavors to accommodate the average person. The belief is that the work environment should fit people and that people should not fit the environment.

Biomechanics is to do with the forces that can be applied by a human body under certain circumstances. The main anatomical considerations in ergonomics are as follows:

- Accessibility of valves, switches, and controls
- Size, shape, and comfort of seats and backrests
- Height and angle of tables and working countertops

An ergonomic checklist for seats would include the following:

- Availability of seats
- Condition of seats
- Height and tilt
- Backrest and lumbar support
- Foot and arm support
- General comfort and adjustment

An ergonomic checklist for the accessibility of instruments and other equipment would include the following:

- Accessibility of valves and gauges
- Switches
- Emergency controls
- Sample points
- Foot holes
- Platforms
- Fixed ladders

Physiological Ergonomics

The second main field of ergonomics is physiology, which concerns itself with the effects of the physical environment on the workplace, as well as the expenditure of energy.

Various factors such as lighting, ventilation, and noise affect people in different ways. Ergonomics investigates and identifies the correct temperatures for work areas, and also identifies where workers are exposed to vibration and other hazards. Work methods are studied in an effort to help prevent fatigue caused by monotonous tasks, and tasks that are too easy are also examined and modified.

An ergonomic checklist for the physiological aspects of ergonomics would include the following:

- Machine vibration
- Muscular strength
- Body tolerance
- Visual acuity
- Hearing ability
- Eye/hand coordination

For the handling of material, the following factors would be considered:

- Mechanical aids
- Back support
- Reach distance
- Minimum carrying distance
- Use of mechanical loaders, etc.

Psychological Ergonomics

Humans act as the sensor when they read instruments, check gauges, and react to warning systems. Because of this interaction, gauges must be easy to interpret and should indicate what they represent. Speedometers in motor vehicles move in a clockwise direction as the speed increases. Fuel gauges indicate high when full and low when empty. This interpretation of signals is an important part of ergonomics and ensures that the sensor reads instruments correctly and takes the correct action.

The information given must be clear and unmistakable and, therefore, the positioning, color, and indication on gauges and controls are important to prevent accidents. One normally expects to open a tap by turning it anti-clockwise and to close by turning it clockwise. Electrical knobs are normally turned clockwise to increase the current and anti-clockwise to reduce the current.

An ergonomic checklist for gauges would be as follows:

- Are the gauges properly mounted and displayed?
- Is there any glare present?
- Are warning signals distinct and separate for different messages?
- Do all the similar gauges read the same way?
- Can the dials and gauges be read quickly and accurately?
- Can any changes or differences on a gauge be easily spotted?
- Will the operator know when the gauge is malfunctioning?

SIGNIFICANT PART OF AN ACCIDENT INVESTIGATION

The identification, selection, and proposal of accident prevention remedies in the form of risk control actions is the most significant part of the entire investigation process. Often, with time, the urgency to fix the accident problem fades. The urgency diminishes and it's back to business as usual. The accident and its consequences are forgotten, and sometimes the implementation of measures to prevent a recurrence is not prioritized. The case goes cold as some would say. Accident investigation files should not be closed until all the remedial actions have been implemented, monitored, and followed upon.

MANAGEMENT ACTION PLAN

To prevent the urgency of implementing risk control measures from waning, a management action plan should be compiled. This is done by the investigators once all the causes of the event have been established. Since this action plan specifies which persons must do what work, and within a certain period, it is less likely to be discarded or ignored.

CONCLUSION

The entire accident investigation process will only be worthwhile and effective if the remedial risk control measures recommended by the investigators are effective in fixing the problem. The cost and benefit of risk reduction are considerations when selecting the remedies. The weaknesses or failure of the SMS should be corrected by the proposed remedies, which should be implemented via a formal action plan and follow-up.

24 Action Plan for Risk Control Remedial Measures

ACTION PLAN METHOD

The Plan, Do, Check, Act (PDCA) cycle, also known as the Deming circle, is a four-step model for carrying out change and continuous improvement. Just as a circle has no end, the PDCA cycle can be repeated for continuous Health and Safety Management System (SMS) improvement. This repetitive management method is used in business for the control and continuous improvement of processes. It is also a recommended method to be used for the implementation of recommended risk control remedial measures, arising from accident investigations.

PLAN

The first step involves *determining* the implementation procedure for the recommended risk control measures proposed by the accident investigators. Objectives, processes, and actions necessary to implement the control measures are compiled. Action plans are set, and policies and performance indicators are established for the implementation of the recommendations.

DO

The second step is the *implementation* of the proposed risk control recommendations. This phase is where information is collected, the action plan is implemented, and the risk control recommendation processes are started. Where applicable, training commences, and the revised, modified, or new SMS standards are written, approved, and implemented.

CHECK

The third step is the *inspection, audit,* and *review* stage. This is when inspections and measurements against the standards, which are established to implement the control measures, take place. It involves monitoring, inspection, audit, measurement, and review of the implemented measures.

ACT

The fourth step is when *corrective action* is initiated to rectify deviations from the planned implementation of the risk control measures. Processes are amended,

DOI: 10.1201/9781003220091-28

modified, and improved to create a system of continual improvement. This may entail re-implementation and modification of the weak or missing SMS elements. It may involve retraining and application of processes to ensure their implementation or rectification. Action plans are compiled to rectify weaknesses in the implementation process. Work for the achievement of implementing the controls is delegated, and timelines for completion are established.

REQUIREMENTS OF AN ACTION PLAN

GOAL

An action plan must have a well-defined description of the goal to be achieved. This means it should specify in detail what the end objective of the actions is. Making the workplace safer is too vague a goal to be realistic, so the goals must be specific. For example, the objective is to conduct an ergonomic assessment of worksites A, B, and C by the end of the month.

STEPS AND TASKS

The action plan should specify what needs to be done to accomplish the goal. It should list the steps, actions, and interventions necessary to implement the recommended risk reduction controls. This may entail

- Writing an SMS standard (Program)
- Revising an SMS standard (Program)
- Improving the existing SMS standards (Programs)
- Retraining or re-skilling employees
- Plant, equipment, and machinery repairs or modifications
- Process changes, etc.

TASKS DELEGATED

The different tasks, duties, and functions specified in the action plan must be allocated to certain individuals or departments within the organization. If a gauge needs to be moved from the top of a tank to a lower position, this task must be assigned to a person or department. If a job specification standard must be written, it should be allocated to persons most suited to write it. Action plan tasks that are not assigned to specific individuals will seldom be done.

TIME BOUND

All work and duties allocated by the action plan must have a timeline assigned. Deadlines and milestones must be plotted for the completion of risk control tasks. If tasks are left open ended, they may drag on forever or simply be forgotten. Some actions may be immediate, some short term, some long term, and others ongoing.

RESOURCES NEEDED

Part of the action plan is a list of the resources needed to complete the tasks and actions required by the action plan. These could be in the form of finances required, manpower needed, or other resources. This will also aid the budgeting process should major expenditures be called for.

MEASURE PROGRESS

The implementation of the risk control measures should be monitored to ensure completion. This may involve a physical inspection of the workplace, documentation review, or employee interviews. The implementation progress should be monitored until all the requirements and recommendations of the accident investigation are implemented and are fully functional.

SMART

All action plan goals should be SMART, meaning that they should be

- Specific – The action plan must specify exactly what must be done in detail. It should not be vague or generalized.
- Measurable and manageable – The objective must be measurable and manageable. Unrealistic goals will never be achieved. If it gets measured, it can be managed.
- Achievable and advantageous – The goal must be achievable considering costs and resources. It should be aligned with the organization's objectives and be advantageous to the organization.
- Realistic and result-oriented – Goals must be realistic and results orientated. A goal such as "injury free" sounds nice, is ideal, but is simply not realistic in a workplace setting. A goal such as the holding of safety committee meetings monthly is achievable and realistic.
- Time-related – All objectives should be time-related and have deadline dates for completion. If left open ended, they will not be achieved. The action plan should specify timelines for the tasks at hand, and allocate milestones for the achievement of the goal. All goals should be tangible.

COST-BENEFIT ANALYSIS

The implementation of risk reduction control measures after an accident should be subjected to a cost-benefit analysis. Business is driven by costs, and management will review all expenditures against the benefit to the company. This decision-making process is based purely on the benefit derived from the expenditure.

A cost-benefit analysis considers the degree of risk reduction gained by the risk reduction intervention, the cost of the intervention, and the resultant savings in financial terms. Cost-benefit analysis involves a comparison of the cost of risk reduction measures, with the risk-factor cost of the accidents prevented. In plain language, it considers how much it will cost to prevent an accident, and determines if that is a good financial investment. Many organizations do not consider the risk of incurring accidental losses and tolerate the risk instead.

CONCLUSION

An accident investigation is not complete until all the risk control measures, proposed by the investigation, have been implemented and monitored for effectiveness. This is the most important phase of an investigation and, if not completed successfully, will not prevent similar accidents from occurring.

A structured approach should be taken for the implementation of the proposed risk control measures by using a formal system such as the Plan, Do, Check, Act (PDCA) cycle, which is a four-step model for carrying out change and continuous improvement, such as the implementation of the recommended risk control measures proposed by the accident investigators. The action plan should be carefully thought out and structured and must include actions, responsibilities, and timelines for completion.

Section V

Evaluating the Quality of Accident Investigation Reports

25 Evaluating the Quality of Accident Investigation Reports

INTRODUCTION

Many accident investigations are not done thoroughly. Many miss the point of an investigation and immediately single out employees to blame. This is an easy cop-out for the investigator, but does not identify the real issues and does not do justice to the injuries suffered by the injured workers or the disruption caused by the event.

MEASUREMENT

Management is often not aware of the quality of the accident investigations being done, and therefore does not know if the workplace risks that caused the accident are being treated correctly and efficiently.

What gets measured, gets managed. To indicate to management the quality of the accident investigations, each accident investigation should be evaluated and scored. Scores should be allocated for each segment of the investigation, and a final percentage allocated.

EXAMINING REPORTS

Once an accident investigation is completed and signed off, it should be submitted to the departmental manager for evaluation. Using a checklist, the manager allocates a score to each portion of the accident investigation form. Only by doing this quality check will management know how thorough and effective the accident investigation system at the organization is.

EVALUATION CHECKLIST

The score sheet must correspond with the accident investigation form, and a score should be allocated to the information given on the form.

SCORING THE ACCIDENT INVESTIGATION REPORT FORM

A score sheet should be drawn up to correspond with the investigation form, and points are allocated to each segment, totaling one hundred points.

GENERAL INFORMATION (10 POINTS)

The general information given on the form should be correct and accurate. This would include the correct names of the employees involved, the date and place of the event, the type of event, and a list of witnesses.

RISK ASSESSMENT (5 POINTS)

The risk assessment portion of the form should be correctly filled in, and should rank what *could* have happened, rather than what actually happened. Probability and severity are to be indicated.

DESCRIPTION OF THE EVENT (15 POINTS)

A brief, accurate, and factual description of the accident should be given. Although it may be difficult for the investigator to accurately determine the costs of the losses, some indication should be given as to the amount, even if it is an estimate. The investigator should indicate if the event was an injury, damage accident, a disruption causing event, or if it was something else such as a fire or accidental exposure. If a combination, this should be indicated.

To evaluate the written description of the event, the description should be

- Comprehensive
- Clear
- Factual
- Accurate
- Detailed
- Event type

The scoring of this portion of the form should allocate points to each of the above criteria (Figure 25.1).

Description of the Accident	Maximum Score	Actual Score
Comprehensive	2	
Clear	2	
Factual	2	
Full description	2	
Details	3	
Type of event	4	
TOTAL	15	

FIGURE 25.1   Score sheet for the accident description.

IMMEDIATE CAUSE ANALYSIS (7POINTS)

High-risk Behaviors

All the high-risk behaviors involved in the event should be listed.

High-risk Workplace Conditions

All the high-risk workplace conditions noted at the accident scene should be listed.

ROOT CAUSE ANALYSIS (8 POINTS)

All the personal (human) factors relating to the immediate causes should be listed, as well as the job (workplace) factors. They must be derived from the immediate causes, and the investigator must be able to justify them. Simply writing down root causes because they seem applicable is not acceptable. They must have been derived by a structured process and must relate to the immediate causes.

SKETCHES/PICTURES/DIAGRAMS

These should be clear and give a good idea of the accident site and losses as a result of the accident. They should also be accurate and factual and relate directly to the event.

RISK CONTROL MEASURES (REMEDIES) (30 POINTS)

This is deemed to be the most important part of the investigation process, and management scrutinizing the accident investigation form will want to know if suitable, effective, remedial measures to reduce the risks have been taken. This portion scores more than any other portion of the accident form. Are these recommended risk control measures feasible, cost-effective and will they treat the root causes? Do they comprise an action plan that delegates tasks in line with the recommendations? Have these tasks been given a commencement and a completion date? Is follow-up action documented? (Figure 25.2).

Risk Control Measures to Prevent a Recurrence	Maximum Score	Actual Score
SMS risk control programs	2	
Positive Steps – High-risk behaviors	5	
Positive Steps – High-risk conditions	5	
Root Causes – Job factors	5	
Root Causes – Personal factors	5	
Accountability	2	
Dates for completion	2	
Date completed	2	
Follow up action	2	
TOTAL	30	

FIGURE 25.2 Score sheet for the risk control measures to prevent a recurrence.

DATE OF INVESTIGATION (20 POINTS)

It is imperative to commence the accident investigation as soon as is practical after the event. Any delays could result in vital evidence being lost, being altered, or missed. Witnesses' recollection of the event also fades with time, so it is important to start the process timeously. Because of the importance of when the investigation was started, 20 points are allocated if the investigation started on the same day. Points are reduced for each day an investigation started after the event.

SIGNATURES (5)

Correct signatures on the form are important to indicate the accident has been reviewed and the signatories are in agreement with the findings. Final signatures are only allocated once all the recommendations for risk control have been implemented, and a follow-up has confirmed this. The form is then submitted to the next one up manager until it reaches the highest level of management within the organization (depending on the severity of the event). The sequence of signatures is as follows:

- Investigator
- Supervisor
- Manager
- One up manager
- Executive
- Safety department

The safety department are the final signatories on the form. Their signature signifies that the details of the accident are accurate, that the root causes have been established, and that the risk control measures are appropriate and have been completely implemented. They also verify that all have signed off on the report form.

HAS THE ACCIDENT INVESTIGATION BEEN EFFECTIVE?

Once a final score has been allocated to the investigation report, management will have a good idea as to how thorough and effective the accident investigation has been. Low scores may indicate a need for retraining in accident investigation, or that the investigator did not take the task seriously enough. This measurement will indicate to management what action needs to be taken to improve the quality of accident investigations. It will also indicate which investigators do a thorough and meaningful investigation.

Section VI

Accident Scenarios

26 Accident Scenario One – Boxcor Manufacturing

INTRODUCTION

The scenario, all names, characters, and incidents portrayed in this accident scenario are fictitious. No identification with actual persons (living or deceased), places, buildings, and products are intended or should be inferred. Here follows the description of the situation that led up to the accident.

BACKGROUND

Boxcor Manufacturing is a large manufacturing industry employing in excess of 500 people. Boxcor was founded over 20 years ago and produces metal shelving racks, shelves, and other metal products. It is situated in the rural community of Ferndale.

According to witnesses, on January 14th, a fork-lift truck made a detour around a walkway that was cluttered with offcuts, pallets, and other excess material. As the fork-lift truck maneuvered past the poor housekeeping, the reverse warning device struck a piece of metal upright protruding too far out of a shelving material storage rack. The bracket of the reverse warning device was bent and needed straightening and grinding to enable the warning device to be refitted back onto the fork-lift truck.

The maintenance team already had more work than they could handle, so the supervisor told one of the workers to find someone to grind the bracket as quickly as possible. The fork-lift truck was urgently required by the shipping department.

THE ACCIDENT

The worker took the bracket to Sid the electrician, who was passing by. While Sid was grinding the bracket on a bench grinder, it dug into the grinding wheel which exploded, shooting a piece of the wheel into his eye. Fragments shot through the ceiling, the light fixture, and the radiator of Sid's pickup truck, which was parked at the workshop entrance. In spite of the medical treatment at the hospital in town, Sid lost sight in his eye.

THE EMPLOYEE

All witnesses agree that Sid was an eager, above-average motivated worker. During his first three years with Boxcor, he worked in electrical maintenance and was responsible for repairs to electric motors, starters, and other switchgear. Sid was a fast efficient worker and after finishing his own work would help the other

DOI: 10.1201/9781003220091-32

electricians with their tasks, or he would wander around visiting other departments to see if he could help them.

Before the accident, Sid had been promoted to the electrical maintenance division, which was a significant promotion. Sid was 25, while most electricians at Boxcor Manufacturing were in their late forties.

THE INVESTIGATION

During the investigation, it became obvious that Sid had not been wearing eye protection at the time of the accident. This is against the company and legal rules. Sid had also been talked to for not wearing personal protective equipment (PPE) in the past. He had also been warned about "visiting" other departments. He had been injured in an accident that occurred a week after he had joined Boxcor Manufacturing five years ago.

ACCIDENT SITE INSPECTION

Inspection of the accident site showed that the condition of the grinding machine was poor. The tool rest was about ½ inch (10mm) from the wheel (it should be 1/8th inch) (3mm). The wheel was not rated for the speed of the grinder, and the tool rest could not be adjusted any closer to the wheel. A dirty, scratched pair of safety glasses and a dust-covered face shield (welding) were hanging nearby, and the fixed eye shield on the machine was dirty, grimy, and not in its correct position.

BOXCOR SAFETY

Boxcor Manufacturing had no procedures covering the maintenance and inspection of grinders, but all maintenance people had been trained in the proper installation of wheels and overall care of machines. This is a prime function of the maintenance apprentice.

Boxcor's policy is that safety is first. There is a program of eye and face protection for grinder operators and all maintenance personnel must wear gloves, hard hats, and safety boots. A major program of erecting safety signs has been in operation for six months. There is also an incentive program that awards employees' injury-free performance. Caps, hard hat stickers, and T-shirts are awarded for each injury-free year worked.

AFTER THE ACCIDENT

After the accident, Sid's supervisor phoned the company doctor and had the safety coordinator drive Sid to the local hospital for medical treatment. Because of the confusion after the accident, a stretcher could not be located and there was a delay. Sid was transported to the hospital in the bed of a pickup truck. The supervisor acted as quickly as he could.

Sid's right eye was severely injured, and as a result, the sight in the eye was completely lost. Boxcor's safety record was spoiled by this accident.

INVESTIGATING THE ACCIDENT

Once the site had been made safe and all the necessary precautions had been taken, the investigation could commence. This accident scenario will now be investigated using the *Logical Sequence Accident Investigation Method* as explained in Chapter 14.

The site inspection revealed the following losses:

- Injury to employee
- Injury – Possible permanent partial disability
- Damage to:
  - Warning device bracket
  - Grinding wheel
  - Grinder
  - Ceiling
  - Light fixture
  - Vehicle radiator
- Work stoppage
- Fork-lift truck operation delayed
- Downtime due to delays
- Production delays
- Investigation time and costs

EXPOSURE, IMPACTS, AND ENERGY EXCHANGES

The following exchanges of energy (impacts) were noted:

- The bracket piece jammed between the wheel and the tool rest causing the wheel to explode.
- Fragments contacted Sid's eye.
- Fragments penetrated the ceiling.
- Fragments damaged the light fixture.
- The vehicle parked outside was damaged by a fragment.

HIGH-RISK BEHAVIORS

The high-risk behaviors were as follows:

- Using unsafe equipment – The grinder was in an unsafe condition.
- Failure to wear PPE – He did not wear eye protection while grinding.
- Improper placement – Detour made by fork-lift truck.
- Unsafe stacking – The material was protruding from the storage rack.
- Working without authority – He was doing a job that was not his normal work.
- Operating at an unsafe speed – He had been told (indirectly) to get the job done in a hurry.
- Unsafe positioning – Transporting an injured employee in the back of a pickup truck.

ROOT CAUSE ANALYSIS – HIGH-RISK BEHAVIOR

To conduct a root cause analysis of this accident, each immediate cause is reviewed by asking, "Why, why, why," to derive the underlying cause of the high-risk action, behavior, or condition.

High-risk behavior – Using an unsafe grinder.

Why? – The grinder had not been maintained, and since the wrong wheel was fitted, the tool rest could not be adjusted any closer to the wheel, leaving a gap that dragged the bracket in and caused the wheel to explode (Root Cause – Job Factor).

Why? – Inadequate maintenance of grinders (Root Cause – Job Factor).

Why? – Unskilled employees (apprentices) do grinder maintenance (Root Cause – Job Factor).

High-risk behavior – Failure to wear PPE.

Why? – Due to a lack of maintenance, the face shield and glasses were dirty and grimy, and in an unusable condition (Root Cause – Job Factor).

Why? – No standards for the maintenance of PPE, no inspection schedule (Root Cause – Job Factor).

Why? – Weak discipline system as previous warnings about not wearing PPE apparently ineffective (Root Cause – Job Factor). (He had been warned about not wearing PPE in the past).

High-risk behavior – Detour of fork-lift truck (?)

Why? – The driver took a decision to bypass the bad housekeeping obstruction. If he had refused, he would more than likely have been reprimanded by the supervisor. If he had climbed down from the truck and moved the obstacles, he would also have been in trouble for doing work that he was not authorized to do (Root cause – Job factor).

Many investigators would say he had committed a high-risk act, but on review, agreed he had done what a person would normally do. The *Reasonable Person* theory applies here. In fact, he did not commit a high-risk act, he did what a reasonable person would do in those circumstances.

Why? – Lack of housekeeping standards or inspections (Root Cause – Job Factor).

High-risk behavior – Transporting an injured employee in the back of a pickup truck.

Why? – There was no stretcher available and probably no first responder (first aider) or first aid equipment on site. There was also no emergency plan which should have included how to transport injured employees to hospital (Root Cause – Job Factor).

Why? – No first aid standards or procedures in place. No training of first responders (first aiders). No emergency equipment available (Root Cause – Job Factor).

High-risk behavior – Working without authority (This was not part of his normal job).

> Why? – Why was Sid in a different department, doing some work other than his own? (Root Cause – Job Factor)
>
> Why? – He was instructed, indirectly, by the supervisor to do the work (Root Cause – Job Factor). (And the supervisor told one of the workers to find someone to grind the bracket as quickly as possible.)
>
> Why? – Issuing incorrect commands and creating conflicting demands (Root Cause – Job Factor).
>
> Why? – Sid was allowed to "visit other departments" indicating his work assignments were insufficient (Root Cause – Job Factor).
>
> Why? – Work assignments not allocated correctly (Root Cause – Job Factor).
>
> Why? – Sid's work was not correctly planned, leaving him too much free time (Root Cause – Job Factor).
>
> Why? – Condoned practice. His habit of "visiting other departments" had been condoned by his leadership (Root Cause – Job Factor).
>
> Why? – He was the youngest of the group of electricians and was probably trying to prove himself by helping others to do their work and "visiting" other departments (Root Cause – Personal Factor).
>
> Why? – He was too young to be promoted to a senior position (Root Cause – Job Factor).
>
> Why? – This created peer pressure (stress) for him to prove himself (Root cause – Personal Factor).
>
> Why? – Visiting and helping others was a condoned practice allowed by his direct supervision, which was as a result of inadequate supervision (Root Cause – Job Factor).

High-risk behavior – Operating at an unsafe speed.

> Why? – He was told the bracket was needed urgently (Root Cause – Personal Factor) (The fork-lift truck was urgently required by the shipping department).
>
> Why? –Improper instruction by the supervisor (Root cause – Job Factor) (…get the job done in a hurry…)

HIGH-RISK WORKPLACE CONDITIONS

The following high-risk workplace environment conditions were noted:

- Poor housekeeping – Fork-lift truck roadway blocked by debris.
- Unsafe stacking and storage – Storage of oversize material in an inadequate storage rack.
- Unguarded machinery – Grinder tool rest gap.
- Inadequate PPE – Dirty face shield, scratched, and dusty eye protection.
- Hazardous environment – Facilities (No first responder [first aider] available).
- Hazardous environment – Facilities (Maintenance team not available).
- Inadequate tools, equipment – First aid equipment lacking, no stretcher.
- Hazardous environment – Facilities (Emergency procedures weak).

ROOT CAUSE ANALYSIS – HIGH-RISK WORKPLACE CONDITIONS

High-risk Workplace Condition – Poor housekeeping (Caused the fork-lift truck to make a detour).

> Why? – Inadequate housekeeping inspection system, or a lack of housekeeping standards (Root Cause – Job Factor).
>
> Why? – Perhaps, the scrap and refuse removal system was inadequate (Root Cause – Job Factor).

High-risk Workplace Condition – Unsafe stacking and storage.

> Why? – The piece of shelving was protruding from the material storage rack because the storage rack was not the correct length for the material (Root Cause – Job Factor).
>
> Why? – Incorrect storage on this rack had been tolerated for a long time (Root Cause – Job Factor).
>
> Why? – Infrequent or inadequate inspection systems did not identify this hazard (Root Cause – Job Factor).

High-risk Workplace Condition – Unguarded machinery (Grinder tool rest gap).

> Why? – Lack of a maintenance plan for grinders (Root Cause – Job Factor).
>
> Why? – Improper wheel had been fitted due to a lack of knowledge (Root Cause – Personal Factor).
>
> Why? – Perhaps poor purchasing led to the wrong size wheel being purchased (Root Cause – Job Factor).

High-risk Workplace Condition – Dirty, scratched face shield.

> Why? – No inspection system for PPE (Root Cause – Job Factor).
>
> Why? – Condoned practice of not wearing face shield while grinding (Root Cause – Job Factor).
>
> Why? – Inadequate supervision, discipline measure ineffective, and PPE standards not available, applied, or both (Root Cause – Job Factor).

High-risk Workplace Condition – No first responder (first aider) available.

> Why? – Inadequate standards on first aid and first responder training (Root Cause – Job Factor).
>
> Why? – No policies or procedures for emergencies (Root Cause – Job Factor).
>
> Why? – Non-availability of first responders (first aiders) (Root Cause – Job Factor).

High-risk Workplace Conditions – Maintenance team not available.

Why? – Incorrect manpower planning and scheduling of maintenance teams, left the organization operating with no maintenance support available (Root Cause – Job Factor).

High-risk Workplace Condition – First aid equipment lacking (no stretcher).

Why? – Inadequate standards on first aid and first aid equipment (Root Cause – Job Factor).
Why? – No policies or procedures for emergencies (Root Cause – Job Factor).

High-risk Workplace Condition – Emergency procedures weak.

Why? – No emergency plans in place. No emergency controller appointed; no drills held. Risk not considered (Root Cause – Job Factor).

INADEQUATE MANAGEMENT CONTROL

The root causes exposed were as a direct result of inadequate controls in the form of a proactive Health and Safety Management System (SMS). Basic elements of the SMS were missing, weak, or non-functional, which resulted in a number of high-risk behaviors being tolerated and high-risk conditions being unrectified.

Weak human resources promotion policies, condoned practices, and inadequate health and safety standards culminated in a major event that cost a worker the sight in one eye. The accident highlighted many weaknesses in Boxcor Manufacturing's approach to health and safety. Typically, the employee would have been blamed for the accident and no other remedial measures would have been forthcoming. In this case, an effective accident investigation clearly identified the many weaknesses and shortcomings that led to the event, which could be rectified to prevent a similar accident in future.

RISK CONTROL REMEDIAL MEASURES TO PREVENT A RECURRENCE

The root cause analysis has uncovered a number of deep-seated problems at Boxcor. Although at first glance, it appears as if Sid was the person solely responsible for the accident, the investigation showed multiple weaknesses in the company's SMS.

Risk control remedial measures to prevent a similar type of accident recurring would include the following recommendations:

1. Employee work allocations must be checked to ensure they have sufficient work for their shift.
2. Supervision must ensure that employees do not enter areas where they normally don't work.
3. Supervision not to issue instructions which can cause unauthorized workers to do tasks that are not their normal duty.

4. All machinery should be placed on a maintenance system. This would include regular checks of grinding machines, including the gap between the wheel and the tool rest, as well as the wheel itself. This would include purchasing standards for grinder wheels, and not allowing unskilled employees (apprentices) to do critical work such as mounting grinder wheels.

5. Correct storage and maintenance for all PPE should be implemented.

6. The inadequate disciplinary procedure to be modified and correctly applied by supervision (merely "talking to an employee" is insufficient).

7. The storage racks are to be modified to suit the materials stored on them.

8. Regular inspections of PPE should be scheduled and carried out.

9. Company SMS standards to be drafted and implemented for:
   • Safety inspections
   • Housekeeping
   • Stacking and storage practices
   • First aid
   • Emergency procedures

10. The human resources department to review the promotion policies and practices to ensure that all employee promotions are scrutinized before they are awarded.

11. The maintenance team work schedules are to be altered to ensure there are maintenance workers during all shifts. If there is a shortage of employees, then this needs to be rectified.

CONCLUSION

By using the *Logical Sequence Accident Investigation Method*, all the aspects of this accident scenario were investigated, and a number of facts were found. Although Sid did commit numerous high-risk behaviors, the existing system allowed him to continue with this behavior.

Numerous immediate causes were determined, and by asking *Why?* many root causes were discovered. The remedial measures are proposals to fix the problems that led to Sid being in a different department, working on an unsafe machine, doing a task that wasn't his normal work, while being pressurized to complete the job in a hurry.

27 Accident Scenario Two – Overhead Electrical Line Accident

INTRODUCTION

The scenario, all names, characters, and incidents portrayed in this accident scenario are fictitious. No identification with actual persons (living or deceased), places, buildings, and products are intended or should be inferred. Here follows the description of the situation that led up to the accident.

BACKGROUND

According to witnesses interviewed, a team of linesmen was erecting an overhead high-tension line spanning a distance of about 1 mile (1.6 km) in a remote area of an industrial park. They were erecting heavy wooden poles into pre-dug holes, then stringing the conductors from pole to pole, and running them through the insulators on top of the poles. This was done from the start point at the transformer to the consumer's connection point in the industrial park.

The terrain was uneven wooded ground. The poles were heavy, 15 inches (0.38 m) in diameter at the bottom, and 12 inches (0.30 m) in diameter at the top, and 27 feet 8 inches (8.4 m) in height. The four conductors were carried by insulators mounted on the sides of the top of the poles.

POLE FALLS

One witness stated that while a linesman, Henry, was up the pole feeding the conductors through the insulators, the pole fell to the ground, slowly at first. Henry loosened his safety belt and jumped to the ground only to be knocked flat by the end of the falling pole. He was badly injured and died a few minutes later. Other witnesses who saw the event agreed on what had happened. They all got the impression that Henry was trying to jump clear of the falling pole, but the pole fell on him when it hit the ground.

The other workers tried to help Henry who was unconscious at the time. A short while later a doctor from the nearby industrial park declared him deceased at the site.

SITE INSPECTION

The site had been left as it was, and the first thing the investigators noticed was the heavy pole was lying on the ground with some conductors still in the insulators. The pole had pulled completely out of the hole. Witnesses said that the pole had created

DOI: 10.1201/9781003220091-33

some sort of a whiplash effect as it came out of the hole and fell onto the hard ground. Investigators also noticed that the hole dug for the pole was in the form of a trench, approximately 19 feet (5.8 m) long and 8 feet (2.4 m) wide. They were told it was 7 feet (2.1 m) deep.

POLE HOLE NOT BACKFILLED

On closer examination, it was apparent that the rectangular-shaped trench dug for the pole had not been completely backfilled or compacted. The pole had only been partially backfilled to a depth of about 5 feet (1.5 m), slightly more than half way. It had not been compacted. Only the side of the trench where the pole had stood was partially backfilled. The trench furthest from the pole's position had fill which sloped down to the end of the trench. It was clear to the investigators that the pole had not been securely supported, as it had not been backfilled, and most of the trench was unfilled.

OTHER POLES INSPECTED

The other poles in the line were also inspected and it was found that some had been almost completely backfilled. Two poles in the line were filled to the same level as the pole that fell. The conductors had already been connected to these poles without incident. Inspectors asked why these poles had not fallen, since a linesman had climbed them, strung the conductors on the insulators, and completed the task safely.

CHANGE OF DIGGING METHOD

Witnesses stated that in the past the holes were always dug by a truck-mounted auger. The auger drilled round holes that allowed the inserted poles to remain upright and secure before final backfilling and compacting. The truck-mounted auger was busy on another site, so the supervisor called in a backhoe to dig the holes. The backhoe was only able to dig trench-like holes and not round holes like an auger.

PARTIAL BACKFILL BY BACKHOE

Once the poles had been inserted into the holes, the backhoe had partially backfilled the holes but had to return to the site office to be loaded onto the flatbed for another job. The backfilling was allocated to the bulldozer.

FINAL BACKFILL BY BULLDOZER

The bulldozer was contacted to come and fill in the partially backfilled holes, and it arrived and awaited radio instructions from the foreman linesman. The foreman was having difficulty with his two-way radio because the battery had run flat, and he could not contact the bulldozer operator. The bulldozer operator got a radio call from the site office to proceed to another job and left the site, without completing the backfilling of the holes.

HELPERS

The four helpers admitted they had helped the victim raise the cables up to the conductors. They were close to the pole and witnessed the entire event. Most of the witnesses said there was a sense of urgency to complete the job as the customer at the industrial park wanted to start operations at his factory.

INVESTIGATING THE ACCIDENT

The accident was investigated using the *Logical Sequence Accident Investigation Method* described in Chapter 14.

This accident resulted in a number of losses:

- Fatality
- The work was stopped
- Legal enquiry and investigation
- The customer was without electricity
- The linesmen were idle for two days during the investigation

EXPOSURE, IMPACTS, AND ENERGY EXCHANGES

The following impacts and exchanges of energy were noted:

- The pole impacted Henry (Struck by).
- The pole impacted the ground (Struck by).

HIGH-RISK BEHAVIORS

The following high-risk behaviors were identified:

- Using equipment incorrectly – Backhoe used instead of an auger.
- Failure to make secure – The pole was not backfilled and secured.
- Taking up an unsafe position – Climbing and working on an unsecured pole.
- Operating at an improper speed – Work commenced before poles were backfilled and compacted.
- Working on unsafe equipment – Hoisting and connecting the conductors to an unsecured pole.
- Failure to warn – His assistants did not caution him that the pole was not backfilled completely.
- Deviation from the standard procedure – Using a backhoe to dig the holes.
- Deviation from the standard procedure – Using a bulldozer to backfill the holes.
- Inadequate communication – Foreman could not contact the bulldozer operator.

ROOT CAUSE ANALYSIS – HIGH-RISK BEHAVIOR

To conduct a root cause analysis of this accident, each immediate cause is reviewed by asking, "Why, why, why," to derive the underlying causes of the action or condition.

High-risk behavior – Using equipment incorrectly (Backhoe used instead of an auger).

> Why? – The auger was unavailable for the job so there was a compromise made and a backhoe was used (Root Cause – Job Factor).
>
> Why? – There was no other way to dig the holes but to deviate from the normally accepted practice (Root Cause – Job Factor).
>
> Why? – The task had not been properly planned (Root Cause – Job Factor).

High-risk behavior – Failure to make secure (The poles were not backfilled nor compacted).

> Why? – On all other jobs the auger had dug neat, round, pole-sized holes which meant that the poles were secure enough in the hole without backfilling (Root Cause – Job Factor).
>
> Why? – The climbing and stringing of conductors could commence in relative safety. (Root Cause – Job Factor). This was a condoned practice.

High-risk behavior – Taking up an unsafe position (Climbing and working on an unsecured pole).

> Why? – In the past, the workers always climbed the poles and strung the conductors once the poles were in the holes, even though they were not completely backfilled (Root Cause – Personal Factor).
>
> Why? – This was a condoned practice due to inadequate leadership (Root Cause – Job Factor).

High-risk behavior – Operating at an improper speed.

> Why? – Witnesses agreed that there was a sense of urgency to complete the job. There was undue pressure to complete the work (Root Cause – Personal Factor).

High-risk behavior – Working on unsafe equipment (Climbing and connecting the conductors to an unsecured pole).

> Why? – This was how it was always done before. The different method of digging and semi-backfilling the holes was a deviation from normal procedures (Root Cause – Job Factor).
>
> Why? – The climbing of the pole once it was upright was common practice to this team (Root Cause – Personal Factor).
>
> Why? – Convenience, which allowed them to carry on despite the poles not being compacted completely (Root Cause – Personal Factor).

Why? – There was a lack of skill in working with poles partially backfilled with a backhoe, as the task method had been changed (Root Cause – Personal Factor).

Why? – Inadequate leadership and supervision, as the supervisor should have recognized the risks of changing the standard methods (Root Cause – Job Factor).

Why? – Inadequate planning to ensure the correct equipment was available to dig and backfill the holes (Root Cause – Job Factor).

High-risk behavior – The deceased assistants did not caution him that the pole was not backfilled completely.

Why? – Failure to warn (Root Cause – Personal Factor).

Why? – They were not trained on backhoe dug holes, or the backfilling and compaction of these types of holes (Root Cause – Personal Factor).

Why? – There was a lack of knowledge of poles being planted in these trench-like holes (Root Cause – Personal Factor).

Why? – Convenience of climbing unsecured poles based on the past work methods. In the past, unsecured poles were wired before they were backfilled (Root Cause – Personal Factor).

Why? – Other poles on site were also partially backfilled and had been successfully wired with no incident. This indicated it was safe to wire all the poles which led to overconfidence (Root Cause – Personal Factor).

High-risk behavior – Using a backhoe to dig the holes.

Why? – Deviation from the standard procedure by supervision (Root Cause – Job Factor).

Why? – The correct equipment was not available (Root Cause – Job Factor).

Why? – Poor job planning forced a change in the standard procedures. Misuse of equipment that was not suited for the job at hand (Root Cause – Job Factor).

High-risk behavior – Using a bulldozer to backfill the holes.

Why? – Deviation from the standard procedure (Root Cause – Job Factor).

Why? – Incorrect instructions were issued by the person in charge of this project (Root Cause – Job Factor).

Why? – Changing from an auger to a backhoe and then to a bulldozer, was a deviation from standard practices (Root Cause – Job Factor).

Why? – This may have been a result of stress caused by the need to connect the customer's electrical supply (Root Cause – Personal Factor).

Why? – Conflicting demands. A further deviation from standard was that the poles were only partially backfilled at the time, and should have been completely backfilled and compacted at the same time (Root Cause – Job Factor).

Why? – The holes were left partially backfilled and the job needed completion (Root Cause – Personal Factor).

Why? – The work crew had to choose between waiting for the bulldozer to return and complete the backfilling, or carry on with the wiring of the poles. This created conflicting demands (Root Cause – Job Factor).

Why? – There was improper motivation as the team was trying to complete the job and meet a deadline (Root Cause – Personal Factor).

High-risk behavior – Foreman could not contact the bulldozer operator.

Why? – Inadequate communication (Root Cause – Job Factor).

Why? – Inadequate maintenance and checking of equipment led to a faulty two-way radio and no communication between the foreman and the bull-dozer (Root Cause – Job Factor).

Why? – Further frustration was created when the bulldozer was recalled to another job (Root Cause – Personal Factor).

HIGH-RISK WORKPLACE CONDITIONS

The following high-risk conditions were identified:

- Hazardous arrangement and inadequate engineering – Change in the normal procedure.
- Incorrect tools or equipment – Backhoe and bulldozer.
- Hazardous environment and insufficiently guarded – Uneven wooded ground and unsecured poles that had not been backfilled.
- Faulty equipment – The two-way radio was faulty.

ROOT CAUSE ANALYSIS – HIGH-RISK WORKPLACE CONDITIONS

High-risk condition – Hazardous arrangement/inadequate engineering.

Why? – There was a change in the normal procedure (Root Cause – Job Factor).

Why? – Inadequate planning meant the correct equipment was not available and a compromise was made to go ahead with the job using what methods and equipment were available (Root Cause – Job Factor).

Why? – The holes dug were the wrong size and shape (Root Cause – Job Factor).

Why? – The backfilling had been started but not completed, thus creating a hazardous situation of unsecured poles (Root Cause – Job Factor).

High-risk condition – A backhoe and bulldozer were used instead of the correct tool, the auger.

Why? – Incorrect tools or equipment were used (Root Cause – Job Factor).

Why? – The auger was not available. The work had to be completed. Poor engineering planning led to inadequate equipment being used (Root Cause – Job Factor).

Why? – A backhoe was used instead of an auger (Root Cause – Job Factor).

Why? – The bulldozer, another piece of equipment never used on this type of job before, was called in to complete the backfilling (Root Cause – Job Factor).

High-risk condition – Unsecured poles that had not been backfilled.

Why? – A hazardous situation was created (Root Cause – Job Factor).

Why? – No risk assessment of the situation had been done. The risk of working on poles that were not secured had not been considered, and no warnings were issued. No hazard identification was undertaken before the poles were climbed (Root Cause – Personal Factor).

High-risk condition – The two-way radio was faulty.

Why? – Faulty equipment on job site (Root Cause – Job Factor).

Why? – Equipment was not checked before the job began. Inadequate maintenance. There was no other method of communicating with the bulldozer operator (Root Cause – Job Factor).

RISK CONTROL REMEDIAL MEASURES TO PREVENT A RECURRENCE

The following risk control remedial measures were recommended:

1. Standard procedures and methods to carry out tasks such as these should never be deviated from, irrespective of the situation or urgency.
2. An auger should always be used to make the holes for the poles, which then should be completely backfilled and compacted before further work commences on the poles.
3. Supervision must oversee the task and conduct on-site risk assessments to ensure the procedure is followed.
4. Adequate communication methods should be available and be in a working condition at all times. A backup radio should be available.
5. Condoned high-risk practices such as climbing unsecured poles should not be allowed or tolerated.
6. Supervision should monitor the progress of the job at all stages and stop the work if it is hazardous.

CONCLUSION

By using the *Logical Sequence Accident Investigation Method*, all the aspects of this accident scenario were investigated, and a number of facts were found. A change in the way the poles were erected led to this fatal accident. Incorrect holes, dug by the backhoe, were of the wrong size and shape, instead of the usual round holes, dug by an auger. The round holes, dug by the auger, allowed the linesmen to climb the erected poles and wire them before they were completely backfilled or compacted. As was the condoned practice, the linesmen proceeded to climb the unsecured poles, erected in oversize holes. One pole fell, killing the linesman. Confusion around the bull-dozer, called in to complete the backfilling, and the poor communication between the foreman and the bulldozer operator also contributed to this adverse event. Changes in the correct procedure should always be subject to a site risk assessment.

Section VII

Conclusion

28 Fixing the Workplace, not the Worker

MISSED OPPORTUNITIES

Around the world, it is reported that there are over 340 million workplace accidents every year. The International Labor Organization estimates that these include 6,000 deaths each day, and the numbers are increasing. Each accident is an opportunity to investigate and determine why the event happened, what caused it, and what can be done to prevent further future accidents.

FIXING THE WORKPLACE, NOT THE WORKER

Instead of using an accident investigation as an opportunity to find and fix the causes, many organizations use the investigation to single out a culprit for the event, and allocate blame and punishment to this worker. While this makes the company look good, a valuable opportunity to identify weaknesses within the organization's management systems is lost. These weaknesses could include fixing the worker rather than *fixing the workplace first*.

PARADIGM SHIFT

Antiquated safety philosophies, based on old and unsubstantiated research, still teach that workers are to blame for the majority of accidents. This has led to the situation where management is absolved of any wrongdoing after an accident. All the blame is put on the worker, who is really a victim of the system. *This safety paradigm needs shifting.*

A CAN OF WORMS

Accidents have multiple causes. Accident investigators cannot simply select the accident causes that suit the organization. Both sides of the coin must be examined and the facts, not the faults, must be clearly and fairly established. This process often opens up a can of worms, and management does not welcome these findings.

THE *LOGICAL SEQUENCE ACCIDENT INVESTIGATION METHOD*

The *Logical Sequence Accident Investigation Method* discussed in this book offers a simple, yet practical approach to an accident investigation. It is easy to follow and facilitates deriving the root causes of the event so that effective risk control measures can be implemented, to prevent a recurrence of a similar accident.

DOI: 10.1201/9781003220091-35

The two accident scenarios in Chapters 26 and 27 give examples of how the investigation is conducted and how immediate and root cause analysis is done. By using the simple but practical *Logical Sequence Accident Investigation Method*, investigators will be in a strong position to deliver an accurate and effective investigation.

CONCLUSION – INVESTIGATORS BEWARE

Although already mentioned, an accident investigation is likely to stir up internal organizational politics which will try to derail the findings of the investigation if it goes beyond blaming the injured parties. The truth revealed by an accident investigation is not easily accepted. Certain parties may duck and dive the findings of the investigation if they are in any way responsible for the accident. There will be finger pointing, shifting of blame, and cop-outs. An effective investigation will open a can of worms. Therefore, investigators beware, the more thorough the investigation, the more intense the pushback.

References

Bateman, Wilson. 2018. *Name, Blame and Shame Game in Incident Investigations.* Incident Management Month 2018 #IMM2018. Incident Investigations: Stop Playing the Blame Game | EHS Blog (pro-sapien.com.) Website 2021.

Benner, Ludwig, Jr. 1978. *Accident Theories and Their Implications for Research.* ACCIDENT THEORIES AND THEIR IMPLICATIONS FOR RESEARCH (iprr.org.)

Canadian Centre for Occupational Health and Safety. 2019a. *Incident Investigation – Physical Evidence.* OSH Answers, Canadian Centre for Occupational Health and Safety (CCOHS), 2019. Reproduced with the permission of CCOHS, 2021. https://www.ccohs.ca/oshanswers/hsprograms/investig.html.

Canadian Centre for Occupational Health and Safety. 2019b. *Incident Investigation – What Should be Done if the Investigation Reveals Human Error?* OSH Answers, Canadian Centre for Occupational Health and Safety (CCOHS), 2019. Reproduced with the permission of CCOHS, 2021. https://www.ccohs.ca/oshanswers/hsprograms/investig.html. CCOHS Website, 2020.

Health and Safety Executive. (HSE.) UK. 1999. *Reducing Error and Influencing Behavior Hsg48.pdf.* (hse.gov.uk.) p.11. (Contains public sector information published by the Health and Safety Executive and licensed under the Open Government License v1.0.)

Health and Safety Executive. (HSE). UK. 2019. *Health and Safety Statistics. Key Figures for Great Britain (2019/20). Health and Safety Statistics.* p.8. Health and Safety Statistics (hse.gov.uk). (Contains public sector information published by the Health and Safety Executive and licensed under the Open Government License v1.0.)

Health and Safety Executive. (HSE). UK. 2004 (a)(b)(c). *Investigating Accidents and Incidents. A Workbook for Employers, Unions, Safety Representatives and Safety Professionals. Hsg245.* Investigating accidents and incidents: A workbook for employers, unions, safety representatives and safety professionals HSG245 (hse.gov.uk) pp.5, 6, 14. (Contains public sector information published by the Health and Safety Executive and licensed under the Open Government License v1.0.)

Health and Safety Executive. (HSE). UK. 2020. *Work-related Stress, Anxiety or Depression Statistics in Great Britain.* p.3. (hse.gov.uk.) (Contains public sector information published by the Health and Safety Executive and licensed under the Open Government License v1.0.)

Heinrich, H. W. 1959. *Industrial Accident Prevention, A Scientific Approach* (4th edition). New York: McGraw-Hill Book Company. Figure 3, p.21.

Hollnagel, Erik. 2021. *The Functional Resonance Accident Model.* https://www.skybrary.aero/index.php/Human_Factors_Analysis_and_Classification_System_(HFACS). Website 2021.

Hoyle, B. 2005. *Fixing the Workplace, not the Worker: A Worker's Guide to Accident Prevention.* Lakewood, CO: Oil, Chemical and Atomic Workers International Union. p.3.

International Labor Organization. (ILO). 2020a. *Encyclopedia of Occupational Health and Safety.* http://www.ilocis.org/documents/chpt56e.htm

International Labor Organization. (ILO). 2020b. *Encyclopedia of Occupational Health and Safety. Theory of Accident Causes – Summary.* https://www.iloencyclopaedia.org/part-viii-12633/accident-prevention/item/894-theory-of-accident-causes.

International Labor Organization. (ILO). 2021. *World Statistics* (ilo.org).

Johnson, Ashley. 2011. *Examining the Foundation: Were Heinrich's Theories Valid? Do They Still Matter?* (safetyandhealthmagazine.com.) Website 2021.

Johnson, Dave. 2018. Industrial Safety and Hygiene News. *Employee Safety Discipline Ain't What It Used to be* | 2018-11-01 | ISHN. ISHN website 2021.

La Duke, Philip. 2011. NSC 2011: *The Top 9 Reasons Workers Don't Report Near Misses* | EHS Today. Website 2011.

McKinnon, Ron C. 2000. *The Cause, Effect, and Control of Accidental Loss, with Accident Investigation Kit. (CECAL).* pp.127, 177, 212. CRC Press, Taylor and Francis Group, 6000 Broken Sound Parkway NW, Suite 300 Boca Raton, FL.

McKinnon, Ron C. 2012. *Safety Management, Near Miss Identification, Recognition and Investigation.* Models 2.4, 2.6, 3.1. pp.27, 39. CRC Press, Taylor and Francis Group, 6000 Broken Sound Parkway NW, Suite 300 Boca Raton, FL.

McKinnon, Ron C. 2017. *Risk-based, Management-led, Audit-driven, Safety Management Systems.* (Figure 2.8) Figures 14.1–14.7 are based on Figure 2.8 A basic accident causation domino sequence. pp.14–17. CRC Press, Taylor and Francis Group, 6000 Broken Sound Parkway NW, Suite 300 Boca Raton, FL.

McKinnon, Ron C. 2019. *Changing Safety's Paradigms.* Maryland: Bernan Press. p.113.

National Occupational Safety Association (NOSA) (1993). *Advanced Questions and Answers on Occupational Safety and Health.* pp.3–8.

National Safety Council (US). *Injury Facts 2019.* 2019 (a, b, c, d). *Work Safety Introduction – Injury Facts* (nsc.org) Permission to reprint/use granted by the National Safety Council © 2021. Website 2021.

Occupational Health and Safety. *OSHA's Top 10 Violations for 2019-- Occupational Health & Safety (ohsonline.com.)*

Occupational Safety and Health Administration (OSHA). 2021. *Recommended Practices on Safety and Health Programs. Management Leadership, Occupational Safety and Health Administration (osha.gov) Safety Management – A Safe Workplace Is Sound Business* | Occupational Safety and Health Administration (osha.gov.)

Reason, James. 2016. *The Swiss Cheese Model of Accident Causation.* https://www.skybrary. aero/index.php/James_Reason_HF_Model#:~:text=The%20Swiss%20Cheese%20 model%20of, gaps%20in%2Dbetween%20each%20slice.

Safeopedia. 2020. *Accident Investigation.* https://www.safeopedia.com/definition/205/accident-investigation-occupational- health-and-safety.

US Bureau of Labor Statistics. 2019. *Fatal Occupational Injuries by Event* (bls.gov.)

Index

Printed in the United States
by Baker & Taylor Publisher Services